Unlocking LaTeX Graphics

A Concise Guide to TikZ/PGF and PGFPLOTS

Tamara G. Kolda

Kolda, Tamara G., Unlocking LaTeX Graphics:
A Concise Guide to TikZ/PGF and PGFPLOTS

latex-graphics.com

Published by MathSci.ai, Dublin, California, 2024

ISBN 979-8-9912295-0-0 (paperback)

Library of Congress Cataloging-in-Publication Data is available upon request.

First Edition, 2024

10 9 8 7 6 5 4 3 2 1

The cover was created entirely by the author using TikZ. (Of course!)

Contents

II PGFPLOTS   49

Preface

TikZ/PGF and PGFPLOTS are incredibly powerful open-source tools for creating high-quality graphics in LaTeX. TikZ/PGF allows the drawing of complex mathematical and technical diagrams, and PGFPLOTS uses the tools of TikZ/PGF to create line, scatter, and bar plots for scientific data visualization. Rather than using external tools to produce your graphics, TikZ graphics compile right along with your LaTeX document. No need to go through a multi-step process to re-render figures when you want to tweak notation or the data changes; instead, they will be regenerated automatically.

Although these packages are powerful, I have found that they are challenging to use. The manual for TikZ/PGF is more than 1,300 pages, and the manual for PGFPLOTS is over 500 pages! While the manuals are comprehensive, they are not necessarily user friendly. I have used the packages for more than a decade in my own technical writing, and these notes are an expansion of cheat sheets that I have developed to speed my own work.

My goal with this book is to boil 1,800 pages down to something that can get you making graphics quickly, focusing on the 10% of TikZ and PGFPLOTS commands that users need 90% of the time. This book is not meant to be a replacement for the manuals, but rather a self-contained reference that enables the reader to do most common operations.

In contrast with the manuals, each of which open with step-by-step tutorials, I have opted for a different approach that attempts to establish fundamentals and build up from there. Once you understand the underlying objects and operations, creating graphics is more efficient.

You can read this book straight through, without needing to jump forward to understand commands that have not yet been introduced. Obviously, there is much more to TikZ/PGF and PGFPLOTS than I cover here, and readers will find that they have a strong base of knowledge on which to build.

You should be able to digest this book in only a few hours. Longer term, it should continue to serve as a handy reference. Each command is listed with all of the relevant key/value options in one place. Once you get familiar with the commands, you may not need any detail other than its option keys.

– Dublin, CA
July 2024

Colors and code conventions

So that you can differentiate what is included in the packages we discuss versus regular LaTeX, code specific to TikZ and PGFPLOTS is color-coded as follows:

- blue — central command or environment
- magenta — TikZ or PGFPLOTS library
- green — key in a key-value pair
- orange — value in a key-value pair *or* style
- purple — path operation
- aqua — math operation (see Section 4.4)
- cyan — PGF number formatting command (see Section 9.2)

In the code and code references, we use the following conventions:

- gray — optional code
- *SlantedText* — variable
- *% gray slanted* — comment
- xy , xy — Mandatory x or y in key, value, or style name
- xy , xy , xy — Optional x or y, with required additional spaces
- Preamble: *Code* — code that goes in document preamble

Acknowledgments and notes.

I'd like to acknowledge the authors of TikZ/PGF and PGFPLOTS for all their hard work in creating these amazing packages! Thanks also to the authors of `tcolorbox`, the LaTeX package that implements many of the visual elements contained in this reference guide. I am grateful to the many contributors on TeX Stack Exchange. Thanks also to my colleagues that have shared their knowledge with me and read advance copies of this book, including Ashley C. Fernandes, Cosmin Safta, and Nico Vervliet. Finally, thanks to my amazing spouse for extensive editing of the final draft and support throughout the process.

Part I: TikZ/PGF

Chapter 1

Enabling TikZ/PGF

TikZ/PGF is a powerful language for creating graphics in LaTeX. PGF is short for Portable Graphics Format. TikZ is a recursive German acronym, short for "TikZ ist *kein* Zeichenprogramm," and is built on top of PGF. The name translates to "TikZ is *not* a drawing program," which is ironic and presumably serves as a warning that you have to program your drawings! To use TikZ, you must include the tikz package in your preamble as shown in Reference 1.1.

Reference 1.1: Enabling TikZ (document preamble)

```
\usepackage{tikz}
```

Within the TikZ package, certain features can only be accessed by including the "library" that contains the feature. Libraries are included in your document via the \usetikzlibrary command, which you will add to the document's preamble. The TikZ libraries we use are listed in Reference 1.2. Whenever these libraries are needed, it is clearly indicated in the examples.

Reference 1.2: Optional TikZ libraries (document preamble)

```
\usetikzlibrary{arrows.meta} % special arrows
\usetikzlibrary{topaths} % curved paths using "to" command
\usetikzlibrary{positioning} % positioning with distance
\usetikzlibrary{shapes.geometric} % additional node shapes
\usetikzlibrary{backgrounds} % "on background layer" scope option
\usetikzlibrary{calc} % coordinate calculations using ($...$)
\usetikzlibrary{math} % math functions
```

1.1   The drawing canvas: `tikzpicture`

The `tikz` package provides the `tikzpicture` environment for drawing figures (discussed here in Part I) or adding plots (discussed in Part II). This is a canvas for your drawing. A `tikzpicture` is sized automatically to its contents. It can be declared anywhere inside the main document, including inside a **center** or **figure** environment. The basic format is shown in Reference 1.3. The environment may have TikZ options inside square brackets, and then the commands for creating the actual picture are used in the main body of the environment.

> **Reference 1.3:** `tikzpicture` **environment**
>
> ```
> \begin{tikzpicture}[TikZOptions]
>   ...
> \end{tikzpicture}
> ```

If we want to specify options for only a subset of commands in a `tikzpicture` environment, we can use a `scope` environment, as shown in Reference 1.4, and declare scope-specific TikZ options inside square brackets.

> **Reference 1.4:** `scope` **environment**
>
> ```
> \begin{tikzpicture}[TikZOptions]
>   ...
>   \begin{scope}[TikZOptions]
>     ...
>   \end{scope}
>   ...
> \end{tikzpicture}
> ```

1.2   Setting global options and styles

A way to specify TikZ options for *all* `tikzpicture` environments is via `\tikzset`. If declared inside an environment such as a **figure**, the options only apply within that scope. This can be useful, for example, for setting the default line thickness (see Section 2.5). These changes are generally made by modifying named *styles*, which are collections of options. Some special named styles serve as defaults; see Section 2.6 for more information.

> **Reference 1.5:** `\tikzset` **for specifying global or scoped options**
>
> ```
> \tikzset{TikZOptions}
> ```

Chapter 2

Paths

The \path command is the main command for drawing lines, circles, grids, etc. It also enables moving around the drawing canvas to position objects relative to one another, define reference points, etc. Section 2.1 gives an overview of the option we will cover, with pointers to more details in the remainder of the chapter.

2.1   Path overview

Many \path operations require specifying coordinates that can be expressed as explicit Cartesian or polar coordinates (Section 2.2), relative coordinates (Section 2.8), or named coordinates (Section 2.11). Paths can also employ *operations* such as those for creating rectangles (Section 2.9), circles/ellipses (Section 2.10), curved paths (Section 2.12), and grids (Section 2.14). Every path command is terminated by a semicolon. We give an overview of the \path command in Reference 2.1. The ⫻ in ⫻ scale means scale, xscale, and yscale are all valid options. At the level of TikZ options, the style every path sets the default path options; see Section 2.6 for more information.

> ⚠ **Remember:** A semicolon is needed at the end of every \path statement.

The path command by itself creates only an invisible path. The draw path modifier tells TikZ to actually draw the path. Likewise, fill fills in the path. The draw and fill path options are special in that they can be options but also have shorthand commands. Many of the TikZ commands operate in this way. Consider the following path aliases:

- \draw is shorthand for \path[draw] (see Section 2.4)
- \fill is shorthand for \path[fill] (see Section 2.4)
- \filldraw is shorthand for \path[fill,draw] (see Section 2.4)
- \coordinate is shorthand for \path coordinate (see Section 2.11)
- \node is shorthand for \path node (see Chapter 3)

The node is a special object in its own right with features that are distinct from those of a path. It is designed as a container for text, even though the text is optional. Nodes can be specified and placed relative to one another without any explicit reference to

---

Reference 2.1: \path (overview)

```
\path[PathOptions] (A) -- (B) ...;
\path[PathOptions] (A) Operation ...;
```

Path option	Description
Color	Color for drawing and filling, defaults to black; see §2.4
draw=Color	Draws the path itself; see §2.4
fill=Color	Fills the path; see §2.4
fill opacity=X	Fill opacity, where X=0.5 is 50%; see §2.4
LineStyle	Line style, e.g., dashed, thick; see §2.5
arrows={...}	Arrows; see §2.7
shorten <=Lmm	Shorten beginning of path, must have units; see §2.7
shorten >=Lmm	Shorten end of path, must have units; see §2.7
ˣʸscale=S	Scale path by a factor of S; see §2.13
shift={(X,Y)}	Shift path by coordinate; see §2.13
xshift=Xcm	Shift path in the x direction, must have units; see §2.13
yshift=Ycm	Shift path in the y direction, must have units; see §2.13
rotate=D	Rotate path by D degrees; see §2.13

Operation	Description
node	Place a node object; see Chapter 3
coordinate	Name a coordinate (command or operation); see §2.11
rectangle	Draw a rectangle; see §2.9
circle	Draw a circle or ellipse; see §2.10
to	Draw a curved path (requires topaths library); see §2.12.1
arc	Draw an arc or a circle; see §2.12.2
grid	Draw a grid; see §2.14

Style	Description
every path	List of options applied to every path; see §2.5

paths or coordinates, so these receive their own chapter (Chapter 3).

Many of the options require some sort of input. If the input contains a comma, as would be the case in manually specifying a coordinate, it must be contained inside curly braces.

> **Remember:** Any argument can be inside curly braces, but they *must* be used if the argument contains a comma, e.g., shift={(X,Y)}.

2.2   Specifying Cartesian and polar coordinates

A coordinate can be specified using (x, y) Cartesian coordinates as (X,Y), or using polar coordinates (D:L) where D is the angle in degrees (not radians) and L is length. Fig. 2.1 shows example Cartesian and polar coordinates; with position $(0, 0)$ labeled at the center. (The TikZ canvas automatically adapts to its contents, so $(0,0)$ has no fixed position in a TikZ picture.) It is also possible to specify 3D (x, y, z) coordinates as `(X,Y,Z)`. Length units can be used but are not required; the default unit of TikZ coordinates is centimeters (cm). To change this default, use the `scale` key as described in Section 2.13.

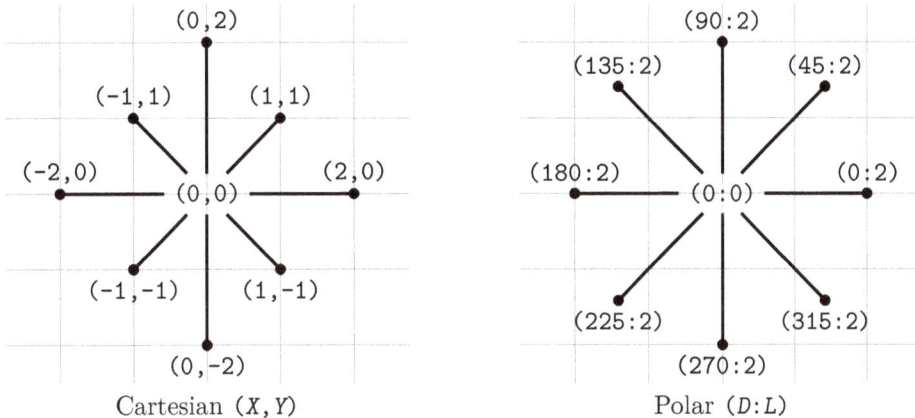

Figure 2.1: Specifying coordinates in TikZ.

⚠ **Remember:** Angle 0 aligns with the positive x-axis at 3 o'clock on the clock face, and angle 90 aligns with the positive y-axis at 12 o'clock.

2.3 Drawing a grid of help lines

Because understanding how paths are rendered onto the underlying coordinate system of the `tikzpicture` is so critical (and potentially confusing), TikZ offers a built-in capability to draw a "help grid" that can be used while creating and debugging your work. Reference 2.2 shows how to draw a light gray unit grid of help lines with the lower-left-hand corner at $(0,0)$ and the upper-right-hand corner at position (X, Y). This is a special case of the `grid` operation, which is discussed in detail in Section 2.14. The `help lines` specification is a predefined *style*, and styles are discussed in Section 2.5.

Reference 2.2: `help lines`

```
\draw[help lines] grid (X,Y);
```

Example 2.1 is an example of drawing a 3 cm × 2 cm grid of help lines, with its lower left corner at position $(0,0)$ and its upper right corner at $(3,2)$.

Example 2.1: Creating a 3 × 2 grid of help lines

```
\begin{tikzpicture}
  \draw[help lines] grid (3,2);
\end{tikzpicture}
```

2.4   Drawing lines and polygons

The most basic \path command you will use is to draw a line, shape, or region and optionally fill the region created by the path. Reference 2.3 provides the structure where (*A*) and (*B*) are coordinates. The draw option specifies to actually draw the line, and the fill option specifies to fill in the region defined by the path. If the draw option is omitted, then the path is invisible but can still be used for other purposes, such as filling or moving around the space.

Reference 2.3: Path drawing and filling

```
\path[PathOptions] (A) -- (B) ...;
\path[PathOptions] (A) -- (B) -- (C) ... -- cycle;
```

Alias	Description
\draw	Alias for \path[draw]
\fill	Alias for \path[fill]
\filldraw	Alias for \path[fill,draw]

Path option	Description
Color	Default color for everything, defaults to black
draw=Color	Draw lines and objects on the path
fill=Color	Fill objects on path
fill opacity=X	Fill opacity, where X=0 is none and X=1 is opaque

---

We next give an example of using the \path command in Example 2.2, recalling that \draw is an alias for \path[draw]. The first command creates a 3×2 help grid (see Section 2.3). The second command draws a line. If we used just \path instead of \draw, no line would be drawn. We specify **red** as the color via the optional input to the draw command; otherwise, the color would default to black. (A list of standard LaTeX colors is provided in Fig. A.2 on page 111.) We use Cartesian coordinates to specify the end points of the line.

Example 2.2: Drawing your first line!

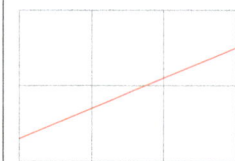

```
\begin{tikzpicture}
  \draw[help lines] grid(3,2);
  \draw[red] (0,0.3) -- (3,1.5);
\end{tikzpicture}
```

Example 2.3 shows usage of the cycle keyword, linking a path back to its start. The first command draws help lines. The second command draws a triangle, using a polar coordinate for the third point and the cycle command to complete the triangle. It's a fine point, but the cycle keyword means that the path has no beginning nor end, in contrast to manually specifying the last point to be the first point.

⚠ **Remember:** The cycle keyword means that the path has no beginning or end.

Example 2.3: Drawing your first cycle! (And using polar coordinates!)

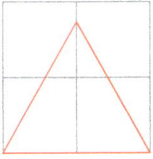

```
\begin{tikzpicture}
  \draw[help lines] grid(2,2);
  \draw[red] (0,0) -- (2,0) -- (60:2) -- cycle;
\end{tikzpicture}
```

In Example 2.4, we declare two styles and use the `fill` keyword. The first command draws help lines. The second command draws a box whose outline is orange and whose fill is specified via `yellow!50` to say it is 50% yellow and the remainder white. Note that the fill is opaque, covering up the help grid lines. (Also, there are easier ways to draw rectangles, discussed later in Section 2.9.) The third command draws a triangle whose specified color is blue, which is used automatically by the `draw` specifier, and `fill=.!10` is shorthand for `fill=blue!10` because it uses the dot to indicate the default color. Even though the path is not a cycle, it is filled as if the last point were connected to the first. In general, the path color can be referenced in the `draw` and `fill` commands via a dot (`.`).

Example 2.4: Filling a path (and using dot-reference for colors)

```
\begin{tikzpicture}
  \draw[help lines] grid(3,3);
  \draw[orange,fill=yellow!50]
    (0,2)--(0,0)--(2,0)--(2,2)--cycle;
  \draw[blue,fill=.!10] (0,3)--(1,1)--(3,2);
\end{tikzpicture}
```

In Example 2.5, we draw a line from (0,1) to (2,3) connecting the two coordinates with `--`, which is referred to as a **line-to** operation. This uses the default color, black. Next, we jump to coordinate (0,0) via a so-called **move-to** operation because there is no `--`. From there, we draw a triangle with additional line-to connections. A move-to has the effect of breaking the path with respect to the `fill` key or `cycle` keyword. In this example, we move back to the origin (0,0) to begin drawing the next set of lines; however, as we'll discuss in Section 2.8, the move-to operation can be used to move relative to the prior coordinate of the path.

Example 2.5: Jumping around with move-to

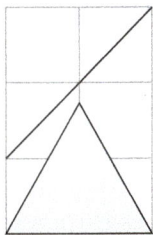

```
\begin{tikzpicture}
  \draw[help lines] grid(2,3);
  \draw[fill=teal!10] (0,1) -- (2,3) (0,0) -- (0:2) --
    (60:2) -- cycle;
\end{tikzpicture}
```

In Example 2.6 on the next page, the `fill opacity` adjusts the opacity, with 1 being opaque (the default) and 0 being completely transparent. The order of the drawing affects what is covered up and what is not. First, we draw grid help lines. Next, we draw a gray triangle, and part of the grid lines is obscured. Finally, we draw a partially opaque violet diamond covering part of the triangle and part of the grid, but these can still be seen because the fill opacity is less than one.

Example 2.6: Invoking semi-transparency

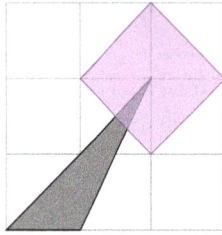

```
\begin{tikzpicture}
  \draw[help lines] grid(3,3);
  \draw[fill=.!50] (0,0)--(1,0)--(2,2)--cycle;
  \draw[violet,fill=.!50,fill opacity=0.6]
      (2,3)--(1,2)--(2,1)--(3,2)--cycle;
\end{tikzpicture}
```

2.5   Line styles

Color is just one modification we can make to a path. We can also change its thickness, change its line pattern, and add arrows. Examples of line thicknesses, patterns, and more are shown in Fig. 2.2. If you are unfamiliar with LaTeX lengths such as "pt," see Fig. A.1 on page 111.

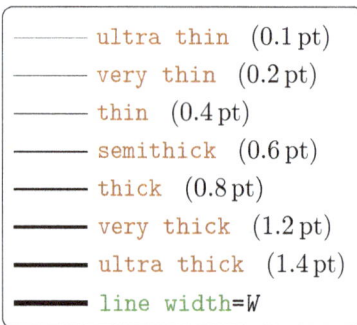

ultra thin (0.1 pt)
very thin (0.2 pt)
thin (0.4 pt)
semithick (0.6 pt)
thick (0.8 pt)
very thick (1.2 pt)
ultra thick (1.4 pt)
line width=W

(a) Line thickness

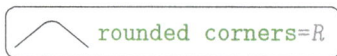

rounded corners=R

(b) Rounded corners

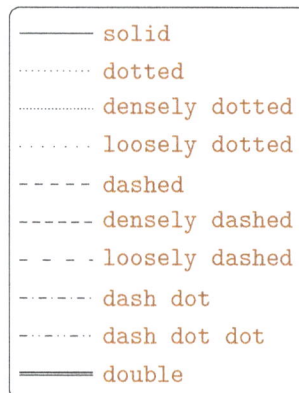

solid
dotted
densely dotted
loosely dotted
dashed
densely dashed
loosely dashed
dash dot
dash dot dot
double

(c) Line pattern

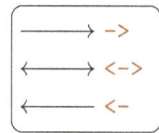

->
<->
<-

(d) Arrows; see §2.7

Figure 2.2: Line and arrow styles. Default is thin and solid.

The modifier rounded corners=R option on a path does exactly as it says: It rounds the corners. The optional argument R specifies the radius of the rounding and defaults to 4 pt.

In Example 2.7, the second draw command draws a winding red line, with rounded corners. Note that rounding the corners has the effect that the line does not always quite touch the designated points; for example, observe the line at the points (3,2) and (1,1) where it does not quite touch the designated points.

Example 2.7: Path with rounded corners

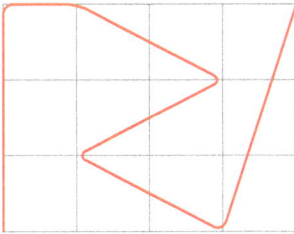

```
\begin{tikzpicture}
  \draw[help lines] grid (4,3);
  \draw[thick,rounded corners,red] (0,0) --
      (0,3) -- (1,3) -- (3,2) -- (1,1) --
      (3,0) -- (4,3);
\end{tikzpicture}
```

2.6   Named styles

PGF has a functionality to build lists of keys and values, which are referred to as *styles* in the context of TikZ. A new named style is declared as *StyleName*/`.style`={*Settings*}. We can also modify a previously created (or system-defined) named style using the *StyleName*/`.append style`={*Settings*} or *StyleName*/`.prefix style`={*Settings*} notation. In the former case, the new style settings are appended to the end of the existing style's declaration. In the latter case, the specified settings are pre-pended. A named style can be declared or modified inside any of the following:

1. \tikzset (outside of tikzpicture, see Reference 1.5 on page 2),
2. arguments to a tikzpicture environment (see Reference 1.3 on page 2), or
3. arguments to a scope environment (see Reference 1.4 on page 2).

The order of declaration matters, and whatever is declared *last* has priority.

Reference 2.4: Defining styles

```
StyleName/.style={Settings}
StyleName/.append style={Settings}
StyleName/.prefix style={Settings}
```

We have already used one named style: help lines, which is defined within TikZ as help lines/`.style`={very thin,gray}. Using the `.append style` operation, we can modify the help lines style to be dashed, as we show in Example 2.8. In this case, we specify the style modification inside the optional arguments for tikzpicture, and so these modifications apply only locally inside this picture.

Example 2.8: Modifying an existing style

```
\begin{tikzpicture}[
    help lines/.append style={dashed}
  ]
  \draw[help lines] grid(3,2);
\end{tikzpicture}
```

We can create our own styles. In Example 2.9, we create two styles, **box** and **diamond**, using \tikzset. Since styles created within a \tikzset are "global," the styles are available from that point forward and could be used in subsequent figures without redeclaration.

Example 2.9: Creating new styles

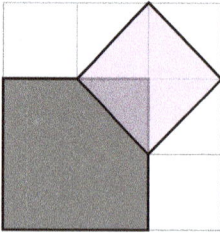

```
\tikzset{
    box/.style={thick,fill=.!50},
    diamond/.style={thick,fill=violet!30, fill opacity=.6}
}
\begin{tikzpicture}
    \draw[help lines] grid(3,3);
    \draw[box] (0,2)--(0,0)--(2,0)--(2,2)--cycle;
    \draw[diamond] (2,3)--(1,2)--(2,1)--(3,2)--cycle;
\end{tikzpicture}
```

Ti*k*Z also has some special predefined styles. The `every path` style is applied to every path, before any other modifications that are passed as arguments. In Example 2.10, we set every path to be `thick` and `dashed` in the Ti*k*Z picture options. Observe that this does not change the thickness of the `help lines` because `every path` is applied *before* `help lines`, which overrides the line thickness to be `very thin`. However, the help lines are still dashed because that is not overridden by any specification in the definition of `help lines`. We use `scope` around the last path command to append `every path` with the style for the diamond.

Example 2.10: Using `every path` and `scope`

```
\begin{tikzpicture}[
    every path/.style={thick,dashed} ]
    \draw[help lines] grid(3,3);
    \draw[fill=.!50] (0,0)--(1,0)--(2,2)--cycle;
    \begin{scope}[every path/.append style={orange,
            fill=.!30, fill opacity=0.6}]
        \draw (2,3)--(1,2)--(2,1)--(3,2)--cycle;
    \end{scope}
\end{tikzpicture}
```

2.7   Arrows

Ti*k*Z boasts an extensive arrow library, and we cover only a few of the options here in Reference 2.5. Arrows can be added to the path via `->` for a single tip at the end, `<-` at the beginning, or `<->` for arrows at both ends. The default arrow tip is `Classical TikZ Rightarrow` and can be changed via `>={ArrowTip}`. Several options for *ArrowTip* are listed in Reference 2.5, such as `Triangle`. A line with two different arrow tips can be specified via the option `arrows={T1-T2}` where *T1* and *T2* are arrow tip names and either can be left blank for a path with a single arrow tip. For example, `arrows={-Triangle}` has an arrow tip of type `Triangle` at its end. There are many further options available in the `arrows.meta` library. Arrows on a path appear only on the first and last segments, and no arrows appear in a cycle. The "shorten" commands are not specific to arrows but are grouped together with the arrows because they enable a path that points at something without actually touching it; see Example 2.12.

In Example 2.11, the second draw command draws a line which has arrows at both ends due to the `<->` option. We set the default arrow style in the `tikzpicture` options via `>=Latex`.

Reference 2.5: arrows

`\path[PathOptions] (A) -- (B) ...;`

Path option	Description
`arrows={T1-T2}`	Arrow tips *T1* at start and *T2* at end; either can be blank
`>={T}`	Set default arrow tip; usually set at higher level
`->`	Shorthand for `arrows={->}`
`<-`	Shorthand for `arrows={<-}`
`<->`	Shorthand for `arrows={<->}`
T1-T2	Shorthand for `arrows={T1-T2}`
`shorten <=`*L*`mm`	Shorten beginning of path by length *L*; needs units
`shorten >=`*L*`mm`	Shorten end of path by length *L*; needs units

Arrow tip	Description
`Classical TikZ Rightarrow[Options]`	$\longrightarrow$ default arrow type
`Straight Barb[Options]`	$\longrightarrow$ requires `arrows.meta`
`Stealth[Options]`	$\longrightarrow$ requires `arrows.meta`
`Latex[Options]`	$\longrightarrow$ requires `arrows.meta`
`Triangle[Options]`	$\longrightarrow$ requires `arrows.meta`

Arrow tip option	Description
Color	Set the color of the arrow (default is path color)
`scale=`*S*	Scale the arrow by *S*

Example 2.11: Setting default arrow style

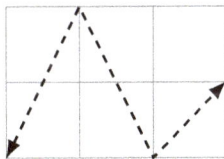

```
Preamble:\usetikzlibrary{arrows.meta}
\begin{tikzpicture}[>=Latex]
  \draw[help lines] grid (3,2);
  \draw[thick,dashed,<->] (0,0) -- (1,2) -- (2,0)
    -- (3,1);
\end{tikzpicture}
```

In Example 2.12, the last command draws an arrow with a `Stealth` tip using `-Stealth`, which is short for `arrows={-Stealth}`. The path is shortened slightly at its end using the `shorten >` key so that the arrow points to the corner of the square without actually touching it.

Example 2.12: Drawing a shortened arrow

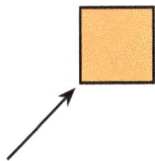

```
Preamble:\usetikzlibrary{arrows.meta}
\begin{tikzpicture}[thick]
  \filldraw[fill=orange!70] (1,1) -- (1,2) -- (2,2) --
    (2,1) -- cycle;
  \draw[-Stealth,shorten >=1mm] (0,0) -- (1,1);
\end{tikzpicture}
```

2.8   Relative coordinates

Up to this point, all of our coordinates have been declared explicitly. However, it's also possible to specify a coordinate by stating its location relative to a reference coordinate, which by default is the prior coordinate in the path. A ++ in front of a coordinate says that the position is *relative* to the reference point (by default the previous coordinate) and updates the reference point to be this newly computed coordinate. A + is identical except that it does not update the reference point.

Using polar coordinates, relative positioning can also be performed by specifying a turn in the path relative to the prior path segment. This is done using the [turn] modifier inside a polar coordinate and is shown in Example 2.15.

Reference 2.6: Relative coordinates

```
\path ... (A) -- ++(B) ...; % Rel move and ref point updated to (B)
\path ... (A) -- +(B) ...; % Rel move but ref point still (A)
\path ... (A) -- (B) -- ([turn]D:R) ...; % Rel turn w.r.t. (A)--(B)
```

In Example 2.13, we draw a cyan diamond with a series of relative points. Then we draw an orange diamond in the same way, and the only thing that has changed (besides the color) is the first point. In this way, it makes it easy to shift a path without needing to change every point.

Example 2.13: Using ++ relative coordinates

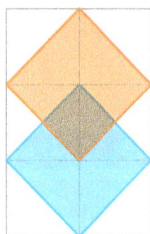

```
\begin{tikzpicture}[thick]
  \draw[help lines] grid(2,3);
  \filldraw[cyan,fill opacity=0.5] (0,1) -- ++(1,1) --
     ++(1,-1) -- ++(-1,-1) -- cycle;
  \filldraw[orange,fill opacity=0.5] (0,2) -- ++(1,1) --
     ++(1,-1) -- ++(-1,-1) -- cycle;
\end{tikzpicture}
```

In Example 2.14, we draw an octagon. The first command draws the help lines. This time all the points are relative to the point $(0,0)$. We draw an octagon by picking out eight evenly spaced points on the unit sphere centered at $(0,0)$. Each point is specified using polar coordinates as +(D:1), where D is the angle in degrees and the + indicates that it is relative to $(0,0)$ and does *not* update the reference point as ++ would. Coordinate $(0,0)$ is only used as a reference point because the first relative point is obtained using a move-to rather than a line-to operation. We round the corners using rounded corners=1pt as a path option.

Example 2.14: Using + relative coordinates

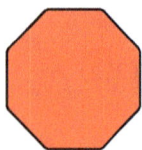

```
\begin{tikzpicture}[thick]
  \draw[fill=red,rounded corners=1pt] (0,0)
    +(22.5:1) -- +(67.5:1) -- +(112.5:1) --
    +(157.5:1) -- +(202.5:1) -- +(247.5:1) --
    +(292.5:1) -- +(337.5:1) -- cycle;
\end{tikzpicture}
```

In Example 2.15, we consider another way to draw an octagon using relative turns. The first line segment is drawn as usual so that we have a reference. Then all subsequent lines are 45-degree turns relative to the previous line segment. The path is completed using `cycle`. We slightly round the edges using `rounded corners=1pt` as a path option.

Example 2.15: Using relative turns

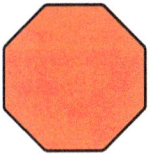

```
\begin{tikzpicture}[thick]
    \draw[fill=red,rounded corners=1pt] (0,0) -- +(0:0.8)
    -- ([turn]45:0.8) -- ([turn]45:0.8) --
    ([turn]45:0.8) -- ([turn]45:0.8) -- ([turn]45:0.8)
    -- ([turn]45:0.8) -- cycle;
\end{tikzpicture}
```

2.9   Drawing rectangles with `rectangle`

The `rectangle` path operation, documented in Reference 2.7, draws a rectangle with opposite corners at **(A)**, which defaults to (0,0), and **(B)**. (Don't confuse this with the `rectangle` node shape; see Section 3.5.) An illustration of the command is shown in Fig. 2.3. The point **(B)** can be in any position relative to **(A)**, and the rectangle will be drawn with respect to those two points. The point **(B)** can be expressed relative to **(A)** by using `++` before the second coordinate. If the two corner points have equal x- or s, then the rectangle reduces to a line. The path continues after a `rectangle` as if it was a move-to operation to the second coordinate. The `rectangle` path operation does not have any options, in contrast to `circle` (see Section 2.10).

Reference 2.7: `rectangle` (path operation)

`\path ... (A) rectangle (B) ...;`

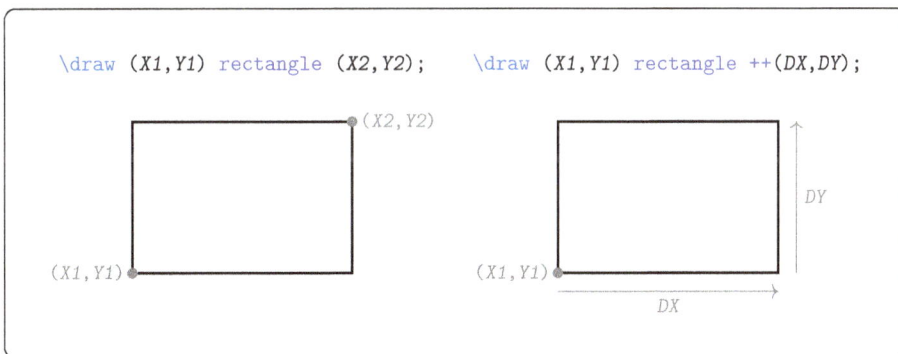

Figure 2.3: Illustration of `rectangle` command

Example 2.16 demonstrates the rectangle command. The orange rectangle does not have a starting point specified, so it defaults to (0,0). If there are no relative points immediately following a rectangle, it does not matter whether its size is specified using `+` or `++`.

Example 2.16: Drawing rectangles with `rectangle`

```
\begin{tikzpicture}[thick]
  \draw[fill=orange] rectangle (3,1);
  \draw[fill=white] (2,1) rectangle +(-2,1);
  \draw[fill=lightgray] (3,2) rectangle (2,1);
  \draw[fill=cyan] (0,2) rectangle ++(1,1);
  \draw[fill=yellow] (3,3) rectangle ++(-2,-1);
\end{tikzpicture}
```

2.10 Drawing circles and ellipses with `circle`

The `circle` path operation draws a circle or ellipse centered at coordinate (C), which defaults to $(0,0)$. (Don't confuse this operation with the `circle` nodes shape; see Section 3.5.) There are several ways of specifying a circle (or ellipse), as shown in Reference 2.8 and illustrated in Fig. 2.4. If no `radius` is specified, it defaults to zero and the circle will be invisible. A circle can be drawn in the middle of a path; the effect is as if there is a move-to operation from the center to itself, meaning that the path resumes from the center of the circle.

Reference 2.8: `circle` (path operation)

```
\path ... (C) circle[CircleOptions] ...;
\path ... (C) circle(Radius) ...;
```

Circle option	Description
`radius=R`	Radius
`x radius=X`	Length in x direction
`y radius=Y`	Length in y direction
`rotate=D`	Rotate by D degrees (only relevant for ellipses)

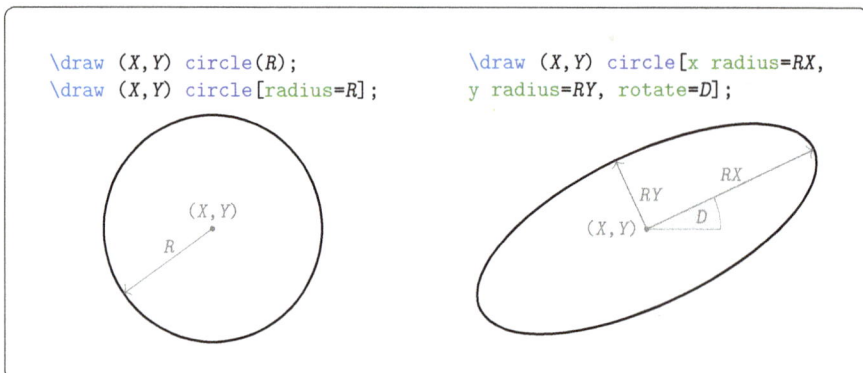

Figure 2.4: Circle path operation

A few examples of circles and ellipses are shown in Example 2.17. The radius of a circle can be specified in two ways. The first is `circle[radius=R]` as is used for the magenta circle. The second is `circle(R)` as is used for the dark gray and black circle. The `circle` path operation can also be used to draw an ellipse by specifying the different x and y radii and any rotation using `circle[x radius=RX, y radius=RY, rotate=D]` as is shown for the white ellipse. We can also specify the circle radius in the path options as we do with the green circles, all drawn with their centers relative to the point $(2, 2)$.

Example 2.17: Drawing circles and ellipses with `circle`

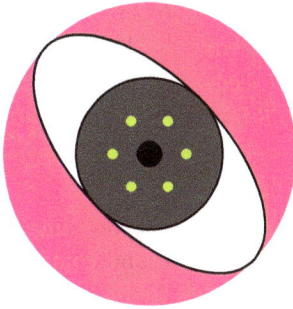

```
\begin{tikzpicture}[thick]
  \fill[magenta] (2,2) circle[radius=2];
  \draw[fill=white] (2,2) circle[x radius=1,
    y radius=2, rotate=45];
  \draw[fill=darkgray] (2,2) circle(1);
  \fill (2,2) circle(5pt);
  \fill[lime,radius=2pt] (2,2)
    +(0:0.5) circle +(60:0.5) circle
    +(120:0.5) circle +(180:0.5) circle
    +(240:0.5) circle +(300:0.5) circle;
\end{tikzpicture}
```

2.11 Named coordinates and coordinate calculations

Named coordinates have a number of uses in TikZ. There are two ways to name a coordinate, with different formatting, as shown in Reference 2.9: (1) using the `\coordinate` command directly, and (2) using `coordinate` in a `\path` command. Names are very flexible, and can be any combination of letters, numbers, and spaces; they cannot contain dots, colons, or semicolons. Additionally, it is possible to name a coordinate at a relative position along a path between two points using the `pos=X` option (see also Section 3.8 for additional placement options).

> ⚠ **Remember:** Names are flexible and can be any combination of letters, numbers, and spaces; however, they cannot contain dots, colons, or semicolons.

Once a few coordinates have been named, they can then be used in "coordinate calculations." For example, if we have named coordinates `foo` and `bar`, then we can reference `(foo |- bar)`, which says to take the x-value from `(foo)` and the y-value from `(bar)`.

The first two "calculations" in Reference 2.9 show how to pull the x-value from coordinate (A) and the y-value from coordinate (B) using the $(A \ |- \ B)$ operation, or vice versa using $(A \ -| \ B)$. The next two calculations show how to compute affine combinations of coordinates (A) and (B), which must be surrounded by $(\$...\$)$ special parentheses. The last calculation computes a relative step of $Gamma$ along the path from (A) to (B); for example, $(\$(A)!0.5!(B)\$)$ is the halfway point.

> ?? **Oddity:** No space is allowed between the asterisk and the opening parenthesis in coordinate calculations that use the `calc` package.

Reference 2.9: Named coordinates and calculations

```
\coordinate (Name) at (A) ...;
\path ... (A) coordinate (Name) ...;
\path ... (A) -- coordinate[pos=X] (Name) (B);
```

Calculation	Description
(A |- B)	Uses x-value from *A* and y-value from *B*
(A -| B)	Uses x-value from *B* and y-value from *A*
($Alpha*(A)+Beta*(B)$)	Affine combination: $\alpha A + \beta B$ (needs calc library)
($Alpha*(A)-Beta*(B)$)	Affine combination: $\alpha A - \beta B$ (needs calc library)
($(A)!Gamma!(B)$)	Move γ proportion from *A* to *B* (needs calc library)

In Example 2.18, we name and reference three of the coordinates and also do some coordinate calculations. The first command draws the help lines. The second command
creates coordinate (A) at position (0.5,1.5) and draws a black dot there. This will
be the corner of the wedge. The third command draws a wedge using calculated coordinates, i.e., the first coordinate is the point (A) plus the polar coordinate (40:3.5)
and the third coordinate is (A) plus the polar coordinate (10:3.5). The fourth command creates the coordinate (B) and draws a black dot. The fifth command draws
an arrow from (B) to (A). The coordinate command creates a coordinate (C) that
will be the center of the circle. The last two commands draw the big and small blue
circles centered at (C).

Example 2.18: Named coordinates and calculations

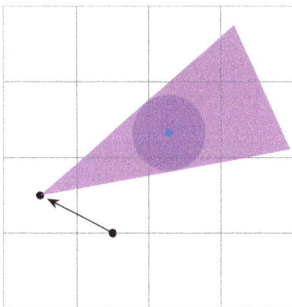

```
Preamble:\usetikzlibrary{arrows.meta,calc}
\begin{tikzpicture}[ >=Stealth,
    wedge/.style={violet,fill opacity=.60},
    bigcirc/.style={blue,fill opacity=0.3}]
\draw[help lines] grid(4,4);
\fill (0.5,1.5) coordinate (A)
    circle(1.5pt);
\fill[wedge] ($(A)+(40:3.5)$) -- (A) --
    ($(A)+(10:3.5)$);
\fill (1.5,1) coordinate (B) circle(1.5pt);
\draw[->,shorten >=1mm] (B) -- (A);
\coordinate (C) at (2.275,2.325);
\fill[bigcirc] (C) circle(0.5cm);
\fill[blue] (C) circle(1.5pt);
\end{tikzpicture}
```

Example 2.19 shows assigning coordinates *along* a path using the pos option when it
draws the line from (B) to (C), labeling a point (L) at 10% along the path and a
point (R) at 90%. Some care is needed in drawing the rectangles. For instance, the red
rectangle starts drawing its southwest corner at (L). The first segment is drawn at the
same angle and then using a series of 90-degree turns. The blue rectangle is similar,
with its southeast corner starting at (R).

Example 2.19: Creating coordinates along a path

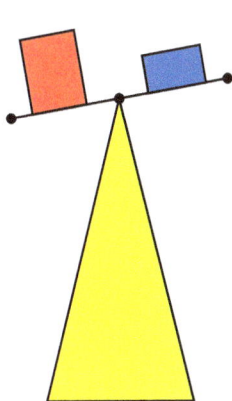

```
\begin{tikzpicture}[thick]
  \draw[fill=yellow] (0,0) -- (0:2) -- (1,4)
    coordinate (A) -- cycle;
  \fill[radius=2pt] (A) circle
    +(10:-1.5) circle coordinate (B)
    +(10:1.5) circle coordinate (C);
  \draw[black] (B) -- (C) coordinate[pos=0.1]
    (L) coordinate[pos=0.9] (R);
  \draw[fill=red] (L) -- ++(10:0.75) --
    ([turn]90:1) -- ([turn]90:0.75) -- cycle;
  \draw[fill=blue] (R) -- ++(10:-0.8) --
    ([turn]270:0.5) -- ([turn]270:0.8) -- cycle;
\end{tikzpicture}
```

Example 2.20 demonstrates using the x- and y-values from different coordinates using the (A |- B) calculation. We draw a blue rectangle, labeling its southwest corner as (BSW) and its northeast corner as (BNE). We next draw a semi-transparent orange rectangle, labeling its southwest corner as (OSW) and its northeast corner as (ONE). We draw the solid arrow using points of the form (BSW |- BNE), i.e., x-value from (BSW) and y-value from (BNE). We draw the dashed arrow using a similar set of points. There are no parentheses around the coordinates, only around the larger expression. If they must be separated, use curly braces, e.g., ({BSW} |- {BNE}).

Example 2.20: Combining x- and y-values from different coordinates

```
\begin{tikzpicture}[thick, fill opacity=0.9,
    arw/.style={->,shorten <=1mm, shorten >=1mm}]
  \filldraw[blue] (0,0) coordinate (BSW) rectangle ++(2,2)
    coordinate (BNE);
  \filldraw[orange] (1,1) coordinate (OSW) rectangle ++(2,2)
    coordinate (ONE);
  \draw[arw] (BSW |- BNE) -- (BSW |- ONE) -- (OSW |- ONE);
  \draw[arw,dashed] (ONE |- OSW) -- (ONE |- BSW) -- (BNE |- BSW);
\end{tikzpicture}
```

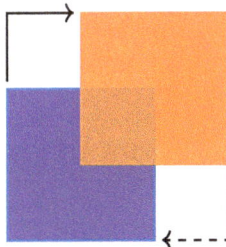

Many other coordinate calculations are possible with TikZ, such as finding intersections of paths, rotations away from a line, projections onto a line, etc. For details, see the TikZ/PGF manual.

2.12   Curved paths

There are various approaches for creating curved paths in TikZ. We discuss two of the
most useful options here: to-paths and arcs. Don't forget that another way to draw a
curved path is to use the `rounded corners` path style; see Section 2.5.

2.12.1   Bending paths with the `topaths` library

The "to path" library requires `\usetikzlibrary{topaths}` in the preamble and enables replacing `--` with `to`. Then we can specify a starting angle (the `out` option) and
an ending angle (the `in` option); see Reference 2.10. The `relative` option means that
the angles are *relative* to the line from (A) to (B), rather than absolute, the default.

Reference 2.10: `to` **(from** `topaths` **library)**

`\path ... (A) to[ToOpts] (B) ...; % Needs topaths library`

To option	Description
`out=ALPHA,in=BETA`	Specify out-angle and in-angle in degrees...
`out=ALPHA,in=BETA,relative`	...relative to straight line between points
`bend left=ALPHA`	Equal to `out=ALPHA,in=180-ALPHA,relative`
`bend right=ALPHA`	Equal to `out=360-ALPHA,in=180+ALPHA,relative`

⚠ **Remember:** Options like `bend left` must be specified using the `to` operator
rather than `--` (line-to).

The `bend left` and `bend right` options are
really useful, but one has to remember how
to interpret "left" and "right." Specifically,
if you are standing on (A) and looking at
(B), then `bend left` goes to the left, whereas
`bend right` goes to the right. In the figure to
the right, the `bend left`=30 curve goes above
the dashed line, while the `bend right`=60
curve goes below.

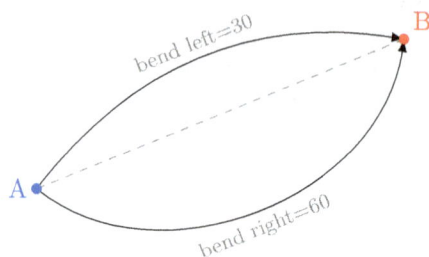

Example 2.21 shows an example of using a to-path with the `bend right` directive. The
`\path` command defines the `coordinate` (A) and then defines (B) and (C) relative to
(A). The `\filldraw` command draws the wedge (B)–(A)–(C) and then arcs from (C)
to (B) using `to[bend right=45]`.

Example 2.21: Drawing a curved line with a `to` **and** `bend right`

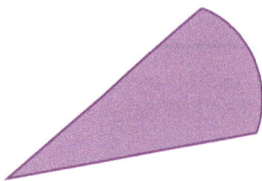

```
Preamble:\usetikzlibrary{topaths}
\begin{tikzpicture}
  \path (0.5,0.5) coordinate (A) +(40:3.5)
    coordinate (B) +(10:3.5) coordinate (C);
  \filldraw[violet, thick, fill=.!50]
    (B)--(A)--(C) to[bend right=45] (B);
\end{tikzpicture}
```

2.12.2 Drawing arcs with `arc`

Another way to get a curved line is to draw an arc along a virtual circle using the `arc` command (see Reference 2.11), but it is a bit tricky to understand the way it works. The arc begins at the point (A), and it is drawn along the arc of a circle of radius R that is centered at the point `($(A)+(-S:R)$)`. We visualize the idea in Fig. 2.5. You can also draw an arc along an ellipse, but we don't cover that option here.

Reference 2.11: `arc` (path operation)

```
\path ... (A) arc[start angle=S, end angle=E, radius=R] ...;
\path ... (A) arc(S:E:R) ...;
\path ... (A) arc[start angle=S, delta angle=D, radius=R] ...;
```

```
\draw (X,Y) arc[start angle=S,end angle=E,radius=R];
\draw (X,Y) arc(S:E:R);
```

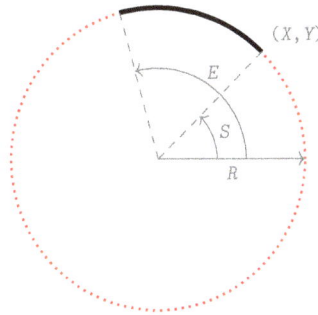

Figure 2.5: Illustration of `arc` operation.

In Example 2.22, we show another way to "cap" the wedge from Example 2.21 with an arc. The `filldraw` command starts at (0.5,0.5), then `-- ++(10:3.5)` draws the first straight line segment, `arc(10:40:3.5)` draws the arced portion, and `-- cycle` finished the wedge.

Example 2.22: Drawing a curved line with arc

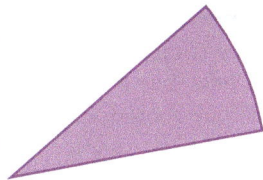

```
\begin{tikzpicture}
  \filldraw[violet,thick,fill=.!50] (0.5,0.5)
    -- ++(10:3.5) arc(10:40:3.5) -- cycle;
\end{tikzpicture}
```

2.13   Path transformations: shift, rotate, scale

We can transform the path, or really the underlying coordinate system for the path, in various ways, as detailed in Reference 2.12. Named coordinates do not transform but rather remain absolute. (Scaling and rotations may not work as expected for nodes along a path; see Section 3.10.)

Reference 2.12: Path transformations: `shift, rotate, scale`

`\path[`*PathOptions*`] ... (A) -- (B) ...;`

Path option	Description
`scale=`*S*	Scale coordinate system by a factor of *S*
`xshift=`*X*`cm`	Shift in the x direction; must have units
`yshift=`*Y*`cm`	Shift in the y direction; must have units
`shift={(`*X*`,`*Y*`)}`	Shift by coordinate; does not need units
`rotate=`*D*	Rotate entire coordinate system by *D* degrees

In Example 2.23, we scale the entire picture, which would normally be $10\,\text{cm} \times 10\,\text{cm}$, to 20% of that size, i.e., $2\,\text{cm} \times 2\,\text{cm}$. The first command draws a red rectangle in the background of size 10×10, which works out to $2\,\text{cm} \times 2\,\text{cm}$ given the scaling of 20%. The second command draws gray rectangles of size 5×5 ($1\,\text{cm} \times 1\,\text{cm}$), the second rectangle starting at the ending corner of the first. The third command draws ten 1×1 ($1\,\text{mm} \times 1\,\text{mm}$) rectangles, one after the other. (We show how to do this more efficiently with a loop in Section 4.5.) The last command draws a grid over everything.

Example 2.23: Scaling an entire picture

```
\begin{tikzpicture}[scale=0.2]
  \fill[red] (0,0) rectangle (10,10);
  \fill[lightgray] (0,10) rectangle ++(5,-5) rectangle
    ++(5,-5);
  \fill[black] (0,10) rectangle ++(1,-1) rectangle
    ++(1,-1) rectangle ++(1,-1) rectangle ++(1,-1)
    rectangle ++(1,-1) rectangle ++(1,-1) rectangle
    ++(1,-1) rectangle ++(1,-1) rectangle ++(1,-1)
    rectangle ++(1,-1);
  \draw (0,0) grid (10,10);
\end{tikzpicture}
```

In Example 2.24, we create a scope that is shifted to the tip of the triangle, point `(A)`, and then rotated 10 degrees. We then draw a line, creating two reference points along the line. We draw a rectangle rooted at each, but we scale the one rooted at `(B)` to 70%. Embedding this inside a scope with the `rotate` option makes it easy to draw the rectangles. For reference, we draw a second help grid inside the rotated scope. The $(0,0)$ position is at the tip of the triangle, and the axes are rotated by 10 degrees.

⚠ **Remember:** Named coordinates do not transform.

Example 2.24: Shift, rotate, and scale

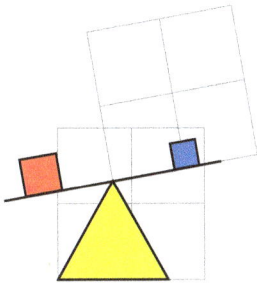

```
\begin{tikzpicture}[thick]
  \draw[help lines] grid(2,2);
  \draw[scale=0.75, fill=yellow] (0,0) --
    (0:2) -- (60:2) coordinate (tip) -- cycle;
  \begin{scope}[shift=(tip),rotate=10]
    \draw[help lines] grid(2,2);
    \draw (-1.5,0) -- coordinate[pos=0.1] (A)
      coordinate[pos=0.9] (B) (1.5,0);
    \draw[fill=red] (A) rectangle ++(0.5,0.5);
    \draw[fill=blue,scale=0.7] (B) rectangle
      ++(-0.5,0.5);
  \end{scope}
\end{tikzpicture}
```

Oddity: The `xshift` and `yshift` options require units, but `shift` does not.

Oddity: The key `xshift` does not have a space following the "x" whereas the circle key `x radius` does.

2.14   Drawing grids with grid

Reference 2.13 describes the `grid` operation, which we have already seen briefly in the context of drawing help lines. The format is similar to `rectangle` (see Section 2.9), but the behavior is slightly different, as we elaborate on below.

Reference 2.13: grid

```
\path ... (A) grid[GridOptions] (B) ...;
\path ... (A) grid[GridOptions] +(D) ...;
```

Grid option	Description
step	Grid size (1 cm)
xstep	Horizontal grid size (1 cm)
ystep	Vertical grid size (1 cm)

Remember: The `grid` command reveals the underlying Cartesian grid; it does not draw a grid rooted at the starting point unless that happens to be a natural grid point.

The best way to think of a grid is as *revealing* the underlying Cartesian grid. Indeed, the most common use of the grid function is to show help lines (see Section 2.3), so this is perhaps one reason for the functionality. Consider Example 2.25. The light gray lines show the Cartesian grid from $(0,0)$ to $(5,5)$. The red grid is a 4×4 "grid" starting at $(0.25, 0.25)$, but it really just forms a bounding box of sorts to display the underlying grid. To get a grid that truly starts at $(0.25, 0.25)$, we have to instead use the shift functionality of the command. This may be counterintuitive.

Example 2.25: Showing grid differences between shift and specifying lower left

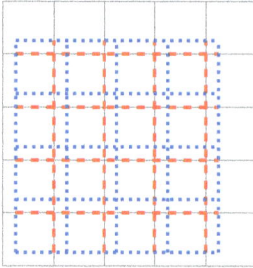

```
\begin{tikzpicture}[very thick,scale=0.7]
  \draw[help lines] grid (5,5);
  \draw[red,dashed] (0.25, 0.25) grid +(4,4);
  \draw[blue,dotted,shift={(0.25,0.25)}]
    grid +(4,4);
\end{tikzpicture}
```

? **Oddity:** A grid with other than unit step size does not always render properly.

Example 2.26 shows an apparent bug in how some grids are displayed when the step is changed from the default of $1\,\mathrm{cm}$. We see that the leftmost grid, which starts at $(0,0)$ is rendered correctly. The second grid, which starts at $(3,0)$, is inexplicably missing its left edge. The third grid, which starts at $(6,0)$, differs in that the step size is unmodified, so the left side is present. The fourth grid shows how to fix the issue by using the shift functionality instead. See also https://tex.stackexchange.com/questions/13834/missing-tikz-grid-borders.

Example 2.26: Oddity of missing grid lines with different steps

```
\begin{tikzpicture}[very thick
  \draw[help lines] grid[step=5mm] (12,2.5);
  \draw[red] (0,0) grid[step=5mm] +(2,2);
  \draw[red] (3,0) grid[step=5mm] +(2,2); % missing left grid line?
  \draw[red] (6,0) grid +(2,2);
  \draw[red,shift={(9,0)}] (0,0) grid[step=5mm] +(2,2);
\end{tikzpicture}
```

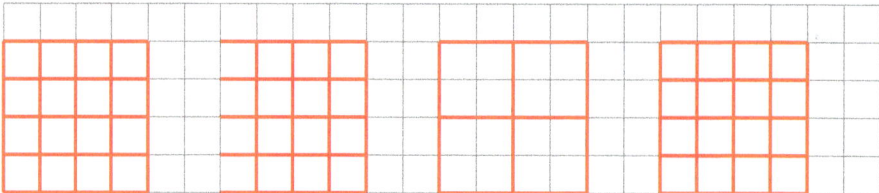

Chapter 3

Nodes

A node is a special path operation that serves as a *container object*. We generally think of nodes as containing text, but they can contain just about anything or be entirely empty. Nodes are designed to conform their shape to whatever they contain. It is possible to create and place nodes without ever directly using a \path command.

Because nodes are so powerful, they receive their own chapter. In Section 3.1 we provide an overview of all the nodes options discussed in this book, and then we provide details on the options in the remainder of the chapter.

3.1   Node overview

The way to create an arbitrary node is the node operation on a path, or \node, which is short for \path node; see Reference 3.1. One purpose of a node is to contain text. The node shapes automatically to contain the text, and certain fields can be used to control text wrapping and alignment. Nodes can alternatively be used to draw shapes along the path, similar to path operations such as rectangle and circle. Any pair of nodes can be connected using an edge; see Section 3.6. A major benefit is that nodes have automatic anchors and advanced facilities for placement; see Sections 3.7 and 3.8. However, determining the size of a node is different and less straightforward. A node can be named and placed like a coordinate; see Section 2.11. At the level of TikZ options, the styles every node and every edge set the defaults, similar to every path as discussed in Section 2.6.

Only a subset of path options are inherited by nodes; for instance, a node inherits the color of a path, but it does not inherit the draw option. A node's size is determined primarily by the text it contains, though we discuss other options below (see Section 3.4). Beyond its ability to contain text, a major feature of a node is reference points (see Section 3.7). Additionally, nodes can be placed relative to one another using the positioning library (see Section 3.8).

Reference 3.1: node (overview)

```
\path ... (A) node[NodeOpts] (Name) {Text} ...;
\node[NodeOpts] (Name) at (A) {Text} ...; % Alternate specification
\path ... (B) edge[EdgeOpts] (C) ...; % Draw edge from (B) to (C)
```

Node option	Description
Color	Color for draw, text, and fill (default black)
fill=Color	Fill node; see §3.2
draw=Color	Draw node edges; see §3.2
LineStyle	Line style options such as thick; see Fig. 2.2
text=Color	Text color; see §3.2
font=Font	Font settings like \itshape, \tiny, etc.; see §3.2
fill opacity=X	Fill and text opacity; see §3.2
text opacity=X	Text opacity, defaults to fill opacity; see §3.2
align=Align	Alignment of text (default left); see §3.3
inner sep=S	Separation of node and text (default 0.3333 em); see §3.4
inner xsep=X	Horizontal separation; see §3.4
inner ysep=Y	Vertical separation; see §3.4
minimum size=S	Minimum size; see §3.4
minimum width=W	Minimum width; see §3.4
minimum height=H	Minimum height; see §3.4
text width=W	Maximum text width; see §3.4
Shape	Shape of node (default rectangle); see §3.5
anchor=Reference	Anchor reference point (default center); see §3.7
Placement	Relative placement (default centered); see §3.8
pos=X	Position along path for $X \in [0,1]$; see §3.9
scale=S	Scale by S; see §3.10
rotate=D	Rotate D degrees around anchor; see §3.10
xyshift=S	Shift (requires units) in specified direction; see §3.10
transform shape	Apply scaling and rotation from path to node; see §3.10

Style	Description
every node	Style of every node (defaults to empty)
every edge	Style of every edge (defaults to draw); see §3.6

Oddity: A node only inherits *some* properties of the path that it's on. It inherits color. It does not inherit draw or fill. It does not inherit rotate or scale unless transform shape is set.

3.2   Node basics

The draw, fill, and fill opacity options are the same as for paths. The text color
is specified by text. The text opacity is inherited from the fill opacity. The
font option specifies the font characteristics like size, family, etc. We cover some basic
node options in Reference 3.2.

Reference 3.2: Basic node options

\path ... (A) node[NodeOptions] (Name) {Text} ...;

Node option	Description
Color	Color for draw, text, and fill (default black)
fill=Color	Fill node (default none)
draw=Color	Draw node edges (default none)
LineStyle	Line style options such as thick; see Fig. 2.2
text=Color	Text color
font=Font	Font settings like \itshape, \tiny, \ttfamily, etc.
fill opacity=X	Node fill, draw, and text opacity (default $X=1$, i.e., opaque)
text opacity=X	Text opacity, defaults to fill opacity

Style	Description
every node	Style of every node (defaults to empty)

?? **?** *Oddity:* Because the text inherits a node's fill opacity by default, you will
likely want to use text opacity=1 whenever the fill opacity is less than one.

In Example 3.1, we show the basics of nodes and edges. We place three nodes in
explicit positions, each of which has a name in parentheses and text in curly braces.
We create a style nd that specifies the style of the node rectangle using options that
were also used for \path in the previous chapter: draw, thick, fill. There are also
two options unique to nodes: text indicates the color of the text, and font indicates
the font characteristics. Observe that modifying font in the (Graphics) node removes
the bold, i.e., it overwrites the prior font setting.

Drawing paths between nodes is somewhat different than drawing paths between points.
For example, the edge from (Precision) to (Clarity) starts at the top edge of
(Precision) and goes to the bottom edge of (Clarity). The positioning of the edge
is determined automatically depending on the relative positions of the nodes. Edges
can be more explicitly controlled, as discussed in Section 3.6.

It is perfectly allowable and oftentimes useful to have *empty nodes*, but don't forget
the curly braces. Example 3.2 demonstrates this. In that example, we also use the
every node style to specify the style of *every* node to be purple and fill, with fill
opacity to 80% so we can see the help lines slightly. Even though the nodes are
empty, they still have some area because inner sep is the default value of 0.3333 em
(see Section 3.4).

Example 3.1: Connected text nodes

```
\begin{tikzpicture}[
    nd/.style={draw,fill=yellow!50,text=blue,font=\bfseries,
      rounded corners},
    arw/.style={thick,shorten >=1mm,shorten <=1mm,->} ]
  \draw[help lines] grid(10,2);
  \node[nd] (Precision) at (1,0.5) {Precision};
  \node[nd] (Clarity) at (3,2)  {Clarity};
  \node[nd,font=\footnotesize] (Graphics) at (6,0.5) {Graphics convey
    complex ideas with simple elegance};
  \draw[arw] (Precision) -- (Clarity);
  \draw[arw] (Graphics) -- (Clarity);
\end{tikzpicture}
```

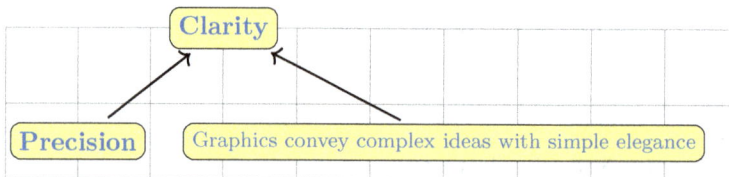

Example 3.2: Connected empty nodes

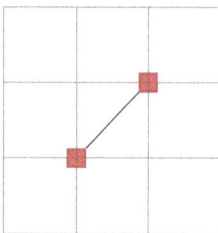

```
\begin{tikzpicture}[every node/.append style=
    {purple,draw,thick,fill,fill opacity=0.8}]
  \draw[help lines] grid(3,3);
  \node (A) at (1,1) {}; % empty node
  \node (B) at (2,2) {}; % empty node
  \draw (A) -- (B);
\end{tikzpicture}
```

⚠ **Remember:** The text of a node is optional, but the curly braces are not!

3.3   Node text width and alignment

Reference 3.3 provides options for the text width and alignment for a node that might have multiple lines of text.

The `text width` option sets the absolute width of the node's text region, even if the text does not need that much space. If the text is wider then the text width, it wraps and the node's height is adjusted automatically to ensure the text is visible. The `align` option controls the positioning of text within the node, either when the text width is specified to be wider than the text contents or when the text wraps onto multiple lines. It is important to note that to have line breaks, either due to wrapping or manually inserted using \\, either `text width` or `align` must be specified. If `text width` is specified, then the alignment defaults to `left`. The options for alignments are given in Fig. 3.1; adding `flush` before `left`, `right`, or `center` minimizes word breaks.

Reference 3.3: Node options `align` and `text width`

`\path ... (A) node[NodeOptions] (Name) {Text} ...;`

Node option	Description
`text width=W`	Fixed text width
`align=Align`	Alignment of text; see Fig. 3.1 for options

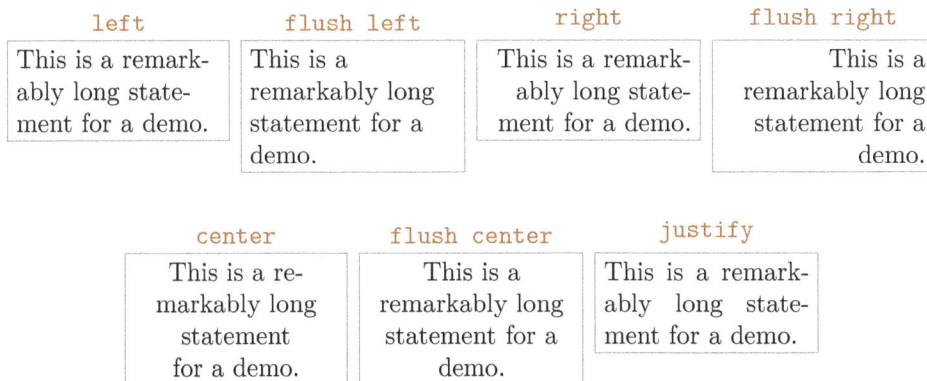

<table>
<tr><td>left</td><td>flush left</td><td>right</td><td>flush right</td></tr>
<tr>
<td>This is a remark-
ably long state-
ment for a demo.</td>
<td>This is a
remarkably long
statement for a
demo.</td>
<td>This is a remark-
ably long state-
ment for a demo.</td>
<td>This is a
remarkably long
statement for a
demo.</td>
</tr>
</table>

<table>
<tr><td>center</td><td>flush center</td><td>justify</td></tr>
<tr>
<td>This is a re-
markably long
statement
for a demo.</td>
<td>This is a
remarkably long
statement for a
demo.</td>
<td>This is a remark-
ably long state-
ment for a demo.</td>
</tr>
</table>

Figure 3.1: Options for `align=Alignment`.

Example 3.3 shows the default behavior, which does not wrap. Just adding a \\ in the node is insufficient to make it wrap; either `text width` or `align` must be specified.

Example 3.3: Line breaks only work if `align` or `text width` is specified

```
\begin{tikzpicture}[nd/.style={draw=gray,very thin},
    ltr/.style={font=\large\bfseries,text=gray}]
  \node[nd] {What a remarkable \\ statement}; % "\\" ignored!
  \node[nd,align=left] at (4,0) {What a\\ remarkable\\ statement};
  \node[nd,text width=4cm] at (8,0) {What a\\ remarkable\\
    statement};
\end{tikzpicture}
```

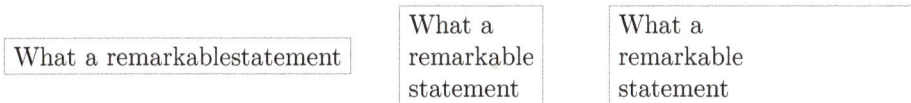

What a remarkablestatement	What a remarkable statement	What a remarkable statement

Oddity: Line breaks (\\) are ignored unless `align` or `text width` is specified.

Example 3.4 revisits Example 3.1, adding a \\ in the long node text and specifying `align=left` in the node style.

Example 3.4: Line breaks and `align`

```
\begin{tikzpicture}[
    every node/.style={draw,thick,fill=yellow!50,font=\bfseries,align=left},
    arw/.style={ultra thick,shorten >=1mm,shorten <=1mm,->}]
  \draw[help lines] grid(7,2);
  \node (Precision) at (1,0.5) {Precision};
  \node at (3,2) (Clarity) {Clarity};
  \node[font=\footnotesize] at (6,0.5) (Graphics) {Graphics convey complex
    ideas\\ with simple elegance};
  \draw[arw] (Precision) -- (Clarity);
  \draw[arw] (Graphics) -- (Clarity);
\end{tikzpicture}
```

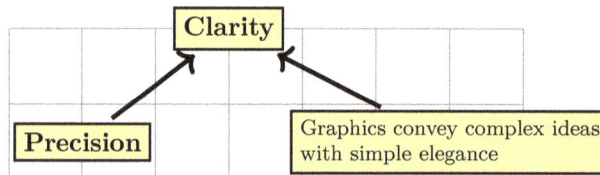

3.4   Node spacing and sizing

Node sizing is automatic, adjusting to fit the text contained therein. However, there are a few options that can be used to modify this, shown in Reference 3.4 and illustrated in Fig. 3.2.

The `inner sep`, `inner xsep`, and `inner ysep` options can be used to modify the width of the whitespace that separates the text from the edge of the node. In the case of a node with no text, the node will still have width and height of at least 2*`inner sep` (unless otherwise overridden); this default behavior was already shown in Example 3.2 on page 26. Additionally, `minimum size`, `minimum width`, and `minimum height` can be used to change the size, and this is especially useful with empty nodes. A maximum size cannot be specified for nodes; the only partial exception is that the `text width` can be specified, which effectively bounds the total width (adding in 2*`inner xsep`), but there is no option to set an upper bound for the height.

Revisiting our prior interconnected node example again in Example 3.5, we set the `text width`=4cm, which forces the text region of all nodes to be 4 cm wide. Observe that the nodes themselves are somewhat wider than 4 cm because of the `inner sep`, which we also enlarge beyond its default so that all the nodes have extra padding. We set `minimum height`=0.7cm, but observe that the long-text node has a height greater than 0.7 cm to accommodate its text. We set `align`=`flush center` to center the text and make it wrap nicely for the long-text node.

We can force empty nodes to be a particular size using options such as `minimum size`, `minimum height`, and `minimum width`. The `inner sep` is what causes an empty node to take up some space by default. Example 3.6, revisiting Example 3.2 on page 26, shows empty nodes that are forced to have a minimum size of 0.5 cm.

Reference 3.4: Node internal spacing and sizing

`\path ... (A) node[NodeOptions] (Name) {Text} ...;`

Node option	Description
`inner sep=S`	Separation of node and text (default 0.3333 em)
`inner xsep=X`	Horizontal separation (default is `inner sep`)
`inner ysep=Y`	Vertical separation (default is `inner sep`)
`minimum size=S`	Minimum size (default is no minimum)
`minimum width=W`	Minimum width (default is `minimum size`)
`minimum height=H`	Minimum height (default is `minimum size`)

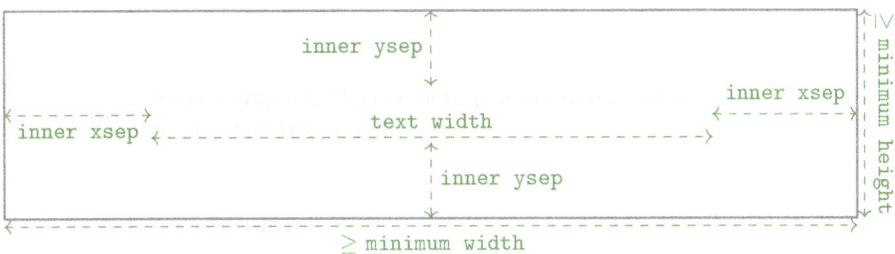

Figure 3.2: Guide to node sizing. If `text width` is not specified, it defaults to the natural width of the text. Using `inner sep` (default 0.3333 em) sets both `inner xsep` and `inner ysep`. Using `minimum size` (default 0) sets both `minimum width` and `minimum height`.

Example 3.5: Sizing nodes using `text width` and `minimum height`

```
\begin{tikzpicture}[
    nd/.style={draw,thick,fill=yellow!50,text width=4cm,
      align=flush center, minimum height=0.7cm, rounded corners},
    arw/.style={ultra thick,shorten >=1mm,shorten <=1mm,->}]
  \draw[help lines] grid(7,3);
  \node[nd] (Precision) at (1,1) {Precision};
  \node[nd] at (3,3) (Clarity) {Clarity};
  \node[nd] at (6,1) (Graphics) {Graphics convey complex ideas with simple
    elegance};
  \draw[arw] (Precision) -- (Clarity);
  \draw[arw] (Graphics) -- (Clarity);
\end{tikzpicture}
```

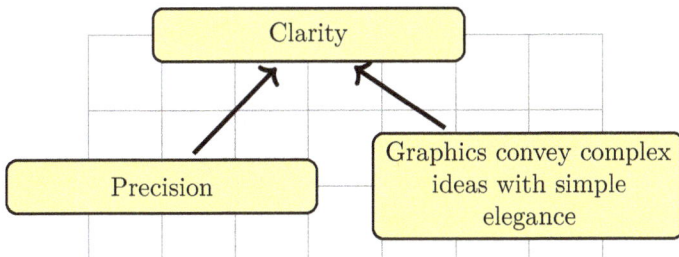

Example 3.6: Sizing empty nodes using `minimum size`

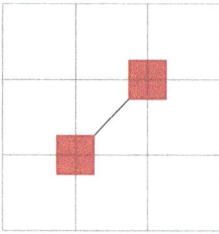

```
\begin{tikzpicture}[every node/.append style=
    {purple, draw, thick, fill, fill opacity=0.8,
      minimum size=0.5cm}]
  \draw[help lines] grid(3,3);
  \path (1,1) node (A) {} (2,2) node (B) {};
  \draw (A) -- (B);
\end{tikzpicture}
```

3.5   Node shapes

There are two main node shapes: `circle` and `rectangle`, which are defined by default. These should not be confused with the path operations of the same name, discussed in Sections 2.9 and 2.10. We contrast the path operations and node shapes in Section 3.11. Other shapes are available in the `shapes.geometric` library, like `diamond`, `ellipse`, `semicircle`, `regular polygon`, `star`, `cylinder`, `cloud`, big arrows called `single arrow` and `double arrow` (like PowerPoint), various call-outs, etc.

In Example 3.7, we have three example node shapes. The default is `rectangle`, used with node "B." The `inner sep` is the same for all three nodes, but results in very different node sizes.

Example 3.7: Circle, square, and diamond node shapes

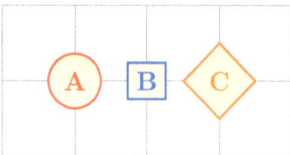

```
Preamble:\usetikzlibrary{shapes.geometric}
\begin{tikzpicture}[nd/.style={draw, thick,
    fill=yellow!25, font=\bfseries}]
  \draw[help lines] grid (4,2);
  \node[red,circle,nd] at (1,1) {A};
  \node[blue,nd] at (2,1) {B};
  \node[orange,diamond,nd] at (3,1) {C};
\end{tikzpicture}
```

> **?** **Oddity:** The `circle` and `rectangle` keywords can specify a node shape *or* a path operation, so be careful not to confuse these with one another.

3.6   Edges

Edges are special paths for connecting coordinates or nodes; see Reference 3.5. To create an edge, we use the `edge[EdgeOpts]` operation within a path in place of the normal line-to (`--`) operation. By default, edges are drawn without specifying the `draw` option. There can be multiple edges in one command, each linking with the first node. Each `edge` can have a different style from the `\path` as well as the other edges in the same command, including changing the color. This is especially useful since you cannot change `path` colors midway. Additionally, an `edge` can follow a node declaration without an explicit `\path` command.

Reference 3.5: Connecting nodes via edge

```
\path ... (A) edge[EdgeOpts] (B) ...; % Edge from (A) to (B)
\node ... (A) {Text} edge[EdgeOpts] (B) ...; % Edge from (A) to (B)
```

Style	Description
every edge	Style of every edge (defaults to draw)

Example 3.8 shows the addition of edges between nodes. Edges can immediately follow the creation of a node, as we demonstrate when creating node B and connecting it to A. Multiple edges can follow a node declaration. Node C has two edges in the same line as its creation; moreover, each has its own style. The edge from C to B is an arrow. The edge from C to A is red. It is possible to declare multiple nodes and edges to each in a single path command, as shown with nodes D and E. Edges inherit the style from the path, so the edges from D to B and E to D are drawn in red and ultra thick. Finally, edges can be drawn between nodes or coordinates that already exist. The last command draws edges from E to C, E to A, and D to E. Observe that each edge is drawn from the last node that is not part of an edge command, so the first edge to A originates with E rather than C, while the second edge to A originates with D rather than E again.

Example 3.8: Connecting nodes with edges

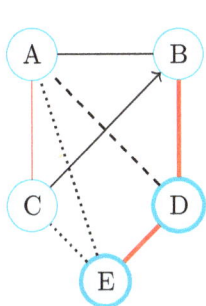

```
\begin{tikzpicture}[every node/.style={circle,
    draw=cyan, fill=cyan!10}]
\path (0,3) node (A){A};
\path (2,3) node (B){B} edge(A);
\path (0,1) node (C){C} edge[->](B) edge[red](A);
\path[draw=red,ultra thick] (2,1) node (D){D}
    edge(B) (1,0) node (E){E} edge(D);
\path[thick] (E) edge[dotted](C) edge[dotted](A)
        (D) edge[dashed](A);
\end{tikzpicture}
```

3.7   Node anchors and shifting

Nodes have *reference points* or *anchor points*, such as north or south west, that can be used (see Reference 3.6) for placement of the node itself or as coordinates in their own right. In addition to the predefined named anchors, *any* integer angle can be a reference point (the point is at the intersection of the node boundary and the line emanating from the center of the node at the specified angle). We show the named anchor points and some example angle anchor points in Fig. 3.3. Nodes can also be shifted via xshift and yshift.

? **?**
 ·?   *Oddity:* Don't forget that south east and other combinations of directions are written as *two* words.

When placing a node at a reference point, the node is centered by default. In this way, the node's *anchor* is its center. By changing the anchor on a node, we can force a

Reference 3.6: Node anchors

`\path ... (A) node[`*`NodeOptions`*`] (`*`Name`*`) {Text} ...;`

Node option	Description
`anchor=`*`Reference`*	Anchor reference point (default `center`); see Fig. 3.3
⟨xy⟩`shift=`*`S`*	Shift (requires units) in specified direction

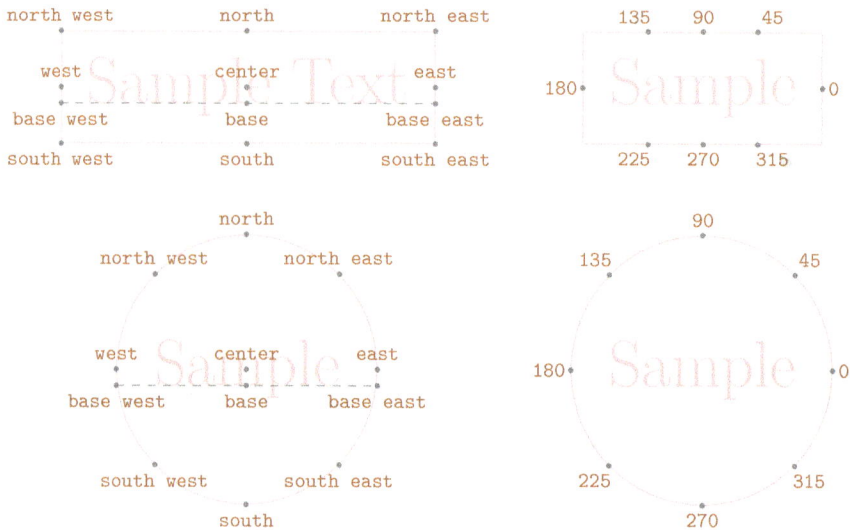

Figure 3.3: Named node reference points and example angle reference points (any angle can be a reference point).

different point on the node's boundary to be placed at the reference point. For example, a node with `anchor=north west` will be positioned such that the top-left corner of the node is placed at the reference point.

In Example 3.9, we use node placement to illustrate a linear algebra equation. (A technique to specify the size using style arguments will be shown in Example 4.16 on page 48.) The first command draws the (A) matrix as a node rectangle. Alternatively, we could have drawn this using the `rectangle` path command, but a path rectangle does not have the reference points of a node rectangle. The second command places a node with an equal sign, starting at (A.`north east`) and moving down 5 mm via `yshift`. We specify `anchor=west` on the node to place it to the right. Now, the equal sign itself is not flush against the (A) node because of the `inner sep` of the node, but the node itself is flush, which we would see if we drew its outline. The next node is (Q), which uses coordinate calculations (see Section 2.11) and sets `anchor=north west` by specifying the `matrix` style. The next node is (R), which we place by starting at (Q.`north east`) and moving right by 1 mm using `xshift`. We modify the matrix (R) to change its height and eliminate its fill. Then we draw a triangle part over top of it, using its reference points. We use `anchor=base` and `yshift=2mm` to align the bottom of the letters slightly above the matrix nodes; if we had instead used `anchor=south`, the protrusion from the letter Q would have made it a bit higher than the A and R.

Example 3.9: Linear algebra equation

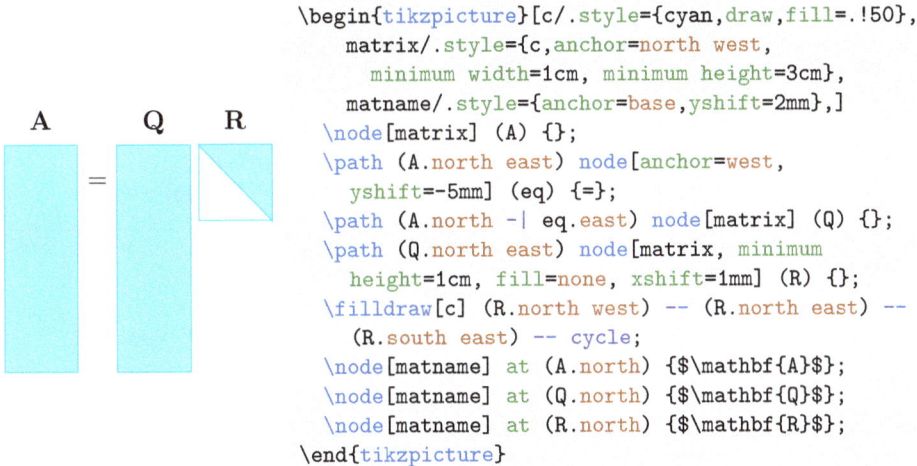

```
\begin{tikzpicture}[c/.style={cyan,draw,fill=.!50},
    matrix/.style={c,anchor=north west,
        minimum width=1cm, minimum height=3cm},
    matname/.style={anchor=base,yshift=2mm},]
\node[matrix] (A) {};
\path (A.north east) node[anchor=west,
    yshift=-5mm] (eq) {=};
\path (A.north -| eq.east) node[matrix] (Q) {};
\path (Q.north east) node[matrix, minimum
    height=1cm, fill=none, xshift=1mm] (R) {};
\filldraw[c] (R.north west) -- (R.north east) --
    (R.south east) -- cycle;
\node[matname] at (A.north) {$\mathbf{A}$};
\node[matname] at (Q.north) {$\mathbf{Q}$};
\node[matname] at (R.north) {$\mathbf{R}$};
\end{tikzpicture}
```

In Example 3.10, we use the node reference points to specify the exact starts and ends of the edges between them.

Example 3.10: Connecting nodes with edges

```
\begin{tikzpicture}[every node/.style={circle,
    draw=cyan, fill=cyan!10}]
\path (0,3) node (A){A} (2,3) node (B){B} (0,1)
    node (C){C} (2,1) node (D){D} (1,0) node (E){E};
\draw (A.east) -- (B.west);
\draw[->] (C.45) -- (B.225);
\draw[red] (C.north) -- (A.south);
\draw[thick,dotted] (C.315) -- (E.135)
    (E.110) -- (A.290);
\draw[ultra thick,red] (D.north) -- (B.south)
    (E.north east) -- (D.south west);
\draw[thick,dashed] (D.135) -- (A.south east);
\end{tikzpicture}
```

3.8   Node placement with respect to other nodes

One nice thing about nodes is that they can be placed relative to one another. There are two ways to specify a placement, detailed in Reference 3.7, with placement locations specified in Fig. 3.4. We describe these two approaches.

The first simple version of placement specifies a choice such as right with respect to the most recent node or coordinate. This is an alternative to node anchors; for instance, right is equivalent to anchor=west.

The second version of the placement requires the positioning TikZ library and enables specification with a distance; for instance, above right=of A would position the new node above and right of (A.north east). No distance is specified, so it defaults to node distance, which itself defaults to 1cm and 1cm (vertical and horizontal). If the specification were above right=1cm of A, then this is the direct distance, i.e., 1 cm at an angle of 45 degrees. The distances are with respect to the outside of the nodes. The node name that follows the "of" should be specified *without* parentheses.

Reference 3.7: Node placement

```
\path ... (A) node[Placement] ...; % Placement relative to A
\path ... node[Placement=Distance of A] ...; % Needs positioning lib
```

Option	Description
Placement	Relative placement, defaults to centered; see Fig. 3.4
node distance	Default placement distance, defaults to 1cm and 1cm

---

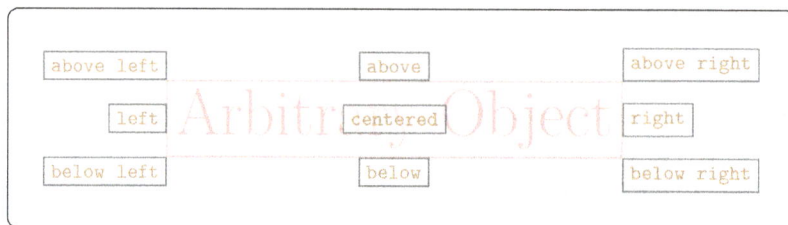

Figure 3.4: Relative placements with respect to an arbitrary object or coordinate.

? *Oddity:* The node name should *not* have parentheses when used in a placement statement.

Example 3.11 shows an example of relative placement. We first place the node A. Node C is above A.north east, which could equivalently be achieved by specifying anchor=south west. Node B is to the right of A.east, so the right specifies anchor=west.

Example 3.11: Arranged nodes

```
\begin{tikzpicture}[every node/.style={draw}]
    \node (A) {A};
    \node[above] at (A.north east) (C) {C};
    \node[right] at (A.east) (B) {B};
\end{tikzpicture}
```

We can do the same thing with the positioning library, as shown in Example 3.12. Observe here that we need not specify A.east in positioning B to the right of A. However, for the more complicated positioning of C, we need to provide the exact coordinate. We could avoid specifying the distance if we add node distance=0mm and 0mm; this is usually most useful as a TikZ option (in the options to tikzpicture or scope or inside \tikzset).

Example 3.12: Arranged nodes with positioning library

```
        Preamble:\usetikzlibrary{positioning}
        \begin{tikzpicture}[every node/.style={draw}]
          \node (A) {A};
          \node[right=0mm of A] (B) {B};
          \node[above=0mm of A.north east] (C) {C};
        \end{tikzpicture}
```

Example 3.13 provides another example of relative node placements. We start by creating node A. Node B is placed above right of A by the default distance (1cm and 1cm). Node C is placed below right of A by 7 mm. Node D is placed above right of C by 10 mm. Node E is placed right of C by the default distance (1 cm). Node F is placed above right of E by 5 mm. Node G is placed above F by the default distance (1 cm).

Example 3.13: Simple graph with relative node placements

```
                    Preamble:\usetikzlibrary{positioning}
                    \begin{tikzpicture}[
                        every node/.style={circle,draw}]
                    \node (A) {A};
                    \node[above right=of A] (B) {B} edge (A);
                    \node[below right=7mm of A] (C) {C} edge (A);
                    \node[above right=10mm of C] (D){D} edge(B)
                      edge(C);
                    \node[right=of C] (E) {E} edge (C) edge (D);
                    \node[above right=5mm of E] (F) {F} edge (D);
                    \node[above=of F] (G) {G} edge (D) edge (F);
                    \end{tikzpicture}
```

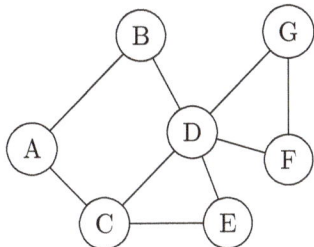

3.9   Node placement along a path

We can define one or more nodes along a path using the syntax in Reference 3.8. The node statement can come before or after the second point with "--" but *must* come before the second point when using "to" (via the topaths library, see Section 2.12.1).

We show two nodes along a path in Example 3.14. Here, the first node is at 3% along the path, and positioned so that it is above right of that point. The second node is 97% along the path, and positioned so that it is above left of that point.

Example 3.14: Placing nodes along a path

```
\begin{tikzpicture}[every node/.style={draw,font=\footnotesize}]
  \draw[thick] (0,0) -- ++(8,0) node[pos=0.03,above right]
    {Compression} node[pos=0.97,above left] {Accuracy};
\end{tikzpicture}
```

| Compression | | Accuracy |

If we tilt the line, as shown in Example 3.15, the nodes are still anchored at the appropriate points along the path, but they are not tilted to match the path.

Reference 3.8: Node placement along path

```
\path ... (A) -- node[Placement,...] (Name) {Text} (B) ...;
\path ... (A) edge node[Placement,...] (Name) {Text} (B) ...;
\path ... (A) to node[Placement,...] (Name) {Text} (B) ...; % topaths
```

Placement	Description
pos=X	Place at position $X \in [0,1]$ along path
at start	Equivalent to pos=0
very near start	Equivalent to pos=0.125
near start	Equivalent to pos=0.25
midway	Equivalent to pos=0.5
near end	Equivalent to pos=0.75
very near end	Equivalent to pos=0.875
at end	Equivalent to pos=1
sloped	Slope node to align with the tangent line at that position
auto	Automatic smart placement next to curve rather than on it

Example 3.15: Placing nodes along a sloped path

```
\begin{tikzpicture}[every node/.style={draw,font=\footnotesize}]
  \draw[thick] (0,0) -- ++(8,0.5) node[pos=0.03,above right]
    {Compression} node[pos=0.97,above left] {Accuracy};
\end{tikzpicture}
```

The sloped node property angles the node to be along the path, or more precisely, to
be parallel with the tangent line at the reference point, as shown in Example 3.16.

Example 3.16: Placing sloped nodes along a sloped path

```
\begin{tikzpicture}[every node/.style={draw,font=\footnotesize,
    sloped}]
  \draw[thick] (0,0) -- ++(8,0.5) node[pos=0.03,above right]
    {Compression} node[pos=0.97,above left] {Accuracy};
\end{tikzpicture}
```

In Example 3.17, we show an example of nodes placed along edges using the auto key
to determine the placement and sloped to align the nodes with the edges.

Example 3.17: Placing nodes automatically along edges using `auto`

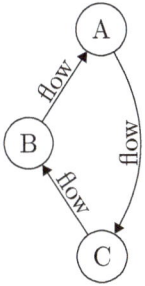

```
Preamble:\usetikzlibrary{arrows.meta}
\begin{tikzpicture}[>=Triangle,
    circ/.style={circle, draw},
    connect/.style={auto, sloped, inner sep=0.1em}]
  \node[circ] at (0,0) (A) {A};
  \node[circ] at (-1,-1.5) (B) {B}
    edge[->] node[connect]{flow} (A) ;
  \node[circ] at (0,-3) (C) {C}
    edge[->] node[connect]{flow} (B);
  \path (A) edge[->,bend left] node[connect]{flow} (C);
\end{tikzpicture}
```

3.10   Scaling and rotation

Nodes can be scaled and rotated; see Reference 3.9. It is important to note that these properties are not, by default, inherited from the path environment. To inherit these properties, use `transform shape`.

Reference 3.9: Node scaling and rotation

`\path ... (C) node[`*NodeOptions*`] (`*Name*`) {`*Text*`} ...;`

Node option	Description
`scale=`*S*	Scale by *S*; not inherited from path!
`rotate=`*D*	Rotate *D* degrees around node's anchor
`transform shape`	Apply scaling and rotation from path to node

It is often useful to rescale an entire picture. However, the nodes do not scale by default. The key `transform shape` a node to apply all transformations of the current path. If a path is rotated and scaled, as in Example 3.18, the `transform shape` key on a node rotates and scales it the the same way. In this case, `sloped` does not have the desired effect, so the "Accuracy" node is neither scaled nor rotated.

Example 3.18: Automatically resized and sloped nodes along a sloped path

```
\begin{tikzpicture}[scale=0.75, every node/.style= {draw}]
  \draw[rotate=3,thick] (0,0) -- ++(8,0)
    node[pos=0.03,above right,transform shape] {Compression}
    node[pos=0.97,above left,sloped] {Accuracy};
\end{tikzpicture}
```

3.11 Node shape versus path operation for `rectangle` and `circle`

The path objects `rectangle` and `circle` and node objects of shapes `rectangle` and `circle` have fundamentally different properties, yet are easy to confuse because they can potentially be swapped for one another. The major difference between the two objects is that nodes can contain text, whereas path objects do not contain text. On the other hand, path objects have fixed sizes, whereas nodes adapt their sizes to the text. Nevertheless, *in either case*, it is possible to mimic the other.

We have the following differences.

- The path objects `rectangle` and `circle` inherit the `draw` and `fill` options from the path, but the node objects of shape `rectangle` and `circle` do not.
- The size of a `rectangle` or `circle` path object is specified by the user, whereas a `rectangle` or `circle` node is sized dynamically depending on its text contents. To get the node object to have a fixed size, we would need to set its `minimum width`, `minimum height`, and/or `text width`, and further ensure that its text is not so large that it exceeds either of those dimensions.
- A node object of shape `rectangle` or `circle` can contain text, but a path object `rectangle` or `circle` does not. Of course, text at the center of a path object can be simulated by adding another node at its center!

There are some further differences unique to each type. Let's start with the `rectangle`.

- The path object `rectangle` is always anchored in the south west, but the node of shape `rectangle` is by default anchored at its center and can specify a different anchor (see Section 3.7).
- Additionally, the path object `rectangle` continues from the north east corner of the path object, whereas the path continues from the anchor of a node of shape `rectangle`.

For a `circle`, we have the following.

- Both the path object `circle` and node of shape `circle` are anchored at their centers, but the node of shape `circle` can specify a different anchor.
- The path object `circle` can also be an ellipse, where as a node must specify a distinct `ellipse` shape.

Chapter 4

Advanced TikZ

4.1 Inline mode

The shorthand `\tikz` shown in Reference 4.1 is equivalent to the `tikzpicture` environment shown in Reference 1.3 on page 2. The commands inside the curly braces must end with semicolons, just as in the `tikzpicture` environment. This `\tikz` shorthand is useful for inline commands, and the special option of note is `baseline`, which specifies how the TikZ canvas should be aligned with the baseline of the surrounding text.

> **Reference 4.1: `\tikz` and `baseline`**
>
> `\tikz[TikZOptions]{TikZCommands}`
>
> | Option | Description |
> |---|---|
> | `baseline=Dpt` | Moves the baseline of TikZ picture by D pt, defaults to 0 pt |
> | `baseline=(A)` | Aligns baseline with coordinate (A) in the TikZ picture |

By default, the TikZ object is created so that the bottom of its bounding box is inline with the text baseline, i.e., the imaginary line that letters sit on. Example 4.1 shows using `\tikz` without any baseline so that we have the default placement and then shifting the TikZ baseline by a negative amount, which has the counterintuitive effect of moving the TikZ object up.

> **Example 4.1: Shifting the TikZ baseline when used inline**
>
> ```
> Compare the default baseline
> \tikz{\filldraw (0,0) -- (0.5,0) circle(1pt) -- (1,0);}
> with a negative shift of -0.5ex
> \tikz[baseline=-0.5ex]{\filldraw (0,0) -- (0.5,0) circle(1pt) -- (1,0);}.
> ```
>
> Compare the default baseline ⸺•⸺ with a negative shift of -0.5ex ⸺•⸺.

?? | *Oddity:* Setting `baseline` to a positive value shifts the picture down, while a negative value shifts the picture up.

It is also possible to choose a coordinate in the picture to align with the baseline of the text. This is especially helpful when inserting text nodes, as we show in Example 4.2.

Example 4.2: Setting the TikZ baseline to a coordinate

Compare `\tikz{\node[fill=yellow]{no baseline};}` to setting the baseline to the `\tikz[baseline=(T.base)]{\node[fill=yellow](T){node baseline};}.`

Compare `no baseline` to setting the baseline to the `node baseline`.

4.2   Current bounding box

TikZ automatically defines various special nodes, the most interesting of which is the current bounding box. This object is a rectangle node whose size is updated as the picture is modified. In Example 4.3, we enlarge the bounding box to create a cyan frame. This has the side effect of changing the bounding box. We enlarge the bounding box again to create a black frame.

Example 4.3: Boxing a picture with current bounding box

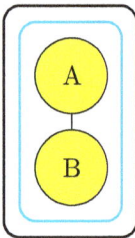

```
Preamble:\usetikzlibrary{positioning,calc}
\begin{tikzpicture}[every node/.style={draw, fill=yellow,
    circle, minimum size=1cm}]
  \path node (A) {A} node[below=2mm of A] (B) {B} edge (A);
  \draw[thick, rounded corners=6pt,cyan]
    ($(current bounding box.south west)+(-0.2,-0.2)$)
    rectangle ($(current bounding box.north east)+(0.2,0.2)$);
  \draw[thick, rounded corners=6pt]
    ($(current bounding box.south west)+(-0.2,-0.2)$)
    rectangle ($(current bounding box.north east)+(0.2,0.2)$);
\end{tikzpicture}
```

4.3   Drawing in the background

By default, TikZ commands within a picture are applied in order, from top to bottom; thus, each command draws on top of everything that has come before. However, it is sometimes useful to be able to draw underneath of what you have already created. The backgrounds library is the simplest option for drawing in the background. It enables the on background layer key for a scope, as shown in Reference 4.2.

Reference 4.2: backgrounds library

```
\begin{scope}[on background layer] % Needs backgrounds library
  ...
\end{scope}
```

We revisit Example 4.3 in Example 4.4 on the facing page, but we want to fill the inner frame with cyan. We cannot draw the frame until after we have drawn the foreground, but if we fill the frame at that stage, it will fill over what we have already drawn.

Instead, we draw and fill the frame in a scope with the option `on background layer` so that the fill of the frame takes place underneath what we have already drawn.

Example 4.4: Drawing in the background using backgrounds library

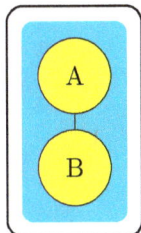

```
Preamble:\usetikzlibrary{positioning,calc,backgrounds}
\begin{tikzpicture}[every node/.style={draw, fill=yellow,
    circle, minimum size=1cm}]
  \path node (A) {A} node[below=2mm of A] (B) {B} edge (A);
  \begin{scope}[on background layer]
    \filldraw[cyan,thick, rounded corners]
      ($(current bounding box.south west)+(-0.2,-0.2)$)
      rectangle
      ($(current bounding box.north east)+(0.2,0.2)$);
  \end{scope}
  \draw[thick, rounded corners=6pt]
    ($(current bounding box.south west)+(-0.2,-0.2)$)
    rectangle ($(current bounding box.north east)+(0.2,0.2)$);
\end{tikzpicture}
```

⚠ **Remember:** Use `\usetikzlibrary{backgrounds}` to make the `on background layer` key available for the `scope` environment.

To create and manipulate multiple layers, see the `\pgfdeclarelayer` and `\pgfsetlayers` commands along with the `pgfonlayer` environment described in the TikZ/PGF manual.

4.4 Math in PGF

Reference 4.3 lists math expressions that are available via the TikZ `math` library, including basic arithmetic and integer calculations. However, we caution users that TikZ mathematical computations are very inexact (see Example 4.5), so be wary.

Math expressions can be used in a variety of situations. The `\pgfmathsetmacro` command stores the result of a math expression in a macro `\A`. (There is a related function `\pgfmathparse` that stores its result by default in the macro `\pgfmathresult`, but it seems that it is convenient to name the variable at the same time as setting the macro.) Parentheses are used to clarify the order of operations or contain a value that may be ambiguous, e.g., `2^(-1)`. We also show how to use an `ifthenelse` statement as a function or use special symbols in the math expression. Math expressions can also be used to define new variables in the context of loops (see Section 4.5), to define special inline variables (see Section 4.6), and to define mathematical functions to be plotted using PGFPLOTS (see Chapter 5).

Math expressions return decimal values, so `\pgfmathsetmacro{\val}{1+1}` sets `\val` to be 2.0 rather than 2. To force it to be an integer without a decimal, use `\pgfmathsetmacro{\val}{int(1+1)}`.

⚠ **Remember:** Use `int` to force an integer rather than a decimal value.

We can calculate the width or height of some text using `width("Text")` or `height("Text")`. As a technical note, you want macros in *Text* to be expanded before its width or

Reference 4.3: Math macros and expressions (needs `math` library)

```
\pgfmathsetmacro{\A}{Expression}
\pgfmathsetmacro{\A}{ifthenelse(Expression,TrueVal,FalseVal)}
\pgfmathsetmacro{\A}{Expression ? TrueVal : FalseVal}
```

Expression	Description
`+,-,*,/,^,!`	Plus, minus, times, divide, power, and factorial
`==,!=,>,<,>=,<=`	Standard comparisons
`||,&&,!`	Standard Boolean operations
`sqrt`	Square root
`abs`, `sign`	Absolute value and sign (`sign(0)=0`)
`round`, `ceil`, `floor`	Standard rounding functions
`int`	Return integer part of number
`div`, `mod`	$div(x,y) = floor(x/y)$, $mod(x,y) = x-y*div(x,y)$
`true`, `false`	Boolean constants
`e,pi`	Constants
`ln,log10,log2`	Logarithmic expressions
`pow,exp`	Power & exponential ($pow(x,y) = x^y$, $exp(x) = e^x$)
`width("Text")`	Calculate width of *Text*
`height("Text")`	Calculate height of *Text*
`rand`	Uniform in $[-1,1]$; seed with `\pgfmathsetseed{Int}`

height is calculated, so macros such as `\emph` or `\small` need the special LaTeX command `\noprotect` in front so that they are expanded before the width or height is calculated.

There is limited support for randomized values. Use `\pgfmathsetseed{Int}` to seed the random number generator, if desired. Then rand produces a uniform value between -1 and 1.

Unfortunately, Example 4.5 shows that even simple TikZ calculations are not accurate.

Example 4.5: Be wary of TikZ math

```
Preamble:\usetikzlibrary{math}
\begin{tikzpicture}
  \pgfmathsetmacro{\t}{10^(-1)}
  \node[rounded corners,draw] (A) {Ti$k$Z math says $10^{-1}=\t$};
\end{tikzpicture}
```

TikZ math says $10^{-1} = 0.09999$

Oddity: Math calculations in TikZ are notoriously inaccurate; use with caution.

Further information on math functions can be found under "Mathematical Expressions" in the TikZ/PGF manual.

4.5   Loops and variables

Reference 4.4 describes the `\foreach` loop command, which can be used inside a `\path` or on its own (and even outside of the TikZ environments). When it's on its own, there can be one or more statements that follow. If multiple statements are used, they must be enclosed in parentheses. The `\foreach` statements can be nested. A single `\foreach` can define multiple variables at each iteration of the loop using slashes between the variables and values. Spaces matter in the loop statements, and extra spaces can sometimes be included in the loop variable definition and cause problems. Of important note for our later coverage, the `\foreach` statement generally does not work with the PGFPLOTS command covered in Part II; see Section 10.14 for further discussion and workarounds.

Reference 4.4: Loops with `foreach`

```
\path ... foreach[LoopOpts] \A in {A1, ...,AN} { ... } ...;
\foreach[LoopOpts] \A in {A1, ...,AN} { ... }
\foreach[LoopOpts] \A/\B in {A1/B1, ...,AN/BN} { ... }
```

Loop option	Description
`count=\C`	Add counter `\C` to the loop
`evaluate=\A as \D`	Evaluate expression for `\A` and store as `\D`
`evaluate=\A as \D using {Exp}`	Evaluate *Exp* (involving `\A`) and store as `\D`
`remember=\A as \F (initially Val)`	Store value from last iteration in `\F`

Example 4.6 re-creates Example 2.23 using `foreach` inside `\path` statements. The second command uses a loop to create two red squares. The `\i` is the loop variable, though we do not reference it anywhere else. The `++(5,-5)` has the effect of moving the starting point for the next rectangle. The third command uses a loop to create ten black squares in a similar fashion.

Example 4.6: Inline foreach example

```
\begin{tikzpicture}[scale=0.2]
\fill[cyan] (0,0) rectangle (10,10);
\fill[red] (0,10) foreach \i in {1,2} {rectangle ++(5,-5)};
\fill (0,10) foreach \i in {1,2,...,10} {rectangle ++(1,-1)};
\draw (0,0) grid (10,10);
\end{tikzpicture}
```

Example 4.7 combines a `\foreach` loop with coordinate calculations (see Section 2.11). The loop sets the variable `\val` and uses that in the calculations for placing the bars on the left and right of node H. In the first iteration of the loop, the expression `($(H.north west)!\val!(H.south west)$)` in the first `\draw` command computes a point that is 10% along the path from the north west to the south west corner of the (H) node. Then the `++(-0.4,0)` moves to the left by 0.4 cm. A node is then placed left of that point. A similar scheme plots the nodes on the right.

Example 4.7: Looping to draw a diagram

```
Preamble:\usepackage{braket} % LaTeX library that provides \ket
Preamble:\usetikzlibrary{calc}
\begin{tikzpicture}
  \node[minimum height=1.5cm,minimum width=1cm,draw] (H) {$H^{\otimes 3}$};
  \foreach \val in {0.1,0.5,0.9} {
    \draw ($(H.north west)!\val!(H.south west)$) --
      ++(-0.4,0) node[left] {$\ket{0}$};
    \draw ($(H.north east)!\val!(H.south east)$) --
      ++(0.4,0) node[right] {$\frac{1}{\sqrt{2}} (\ket{0} + \ket {1})$}; }
\end{tikzpicture}
```

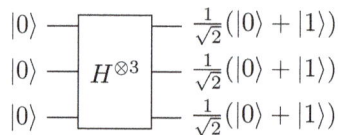

$$|0\rangle \;\text{—}\; \boxed{\phantom{H^{\otimes 3}}} \;\text{—}\; \tfrac{1}{\sqrt{2}}(|0\rangle + |1\rangle)$$
$$|0\rangle \;\text{—}\; \boxed{H^{\otimes 3}} \;\text{—}\; \tfrac{1}{\sqrt{2}}(|0\rangle + |1\rangle)$$
$$|0\rangle \;\text{—}\; \phantom{\boxed{H^{\otimes 3}}} \;\text{—}\; \tfrac{1}{\sqrt{2}}(|0\rangle + |1\rangle)$$

The count option in a foreach loop is demonstrated in Example 4.8. We loop through the letters 'a' to 'z' using the variable \ltr and keep a count in the macro \cnt that we use for positioning the nodes.

Example 4.8: Loop with counting

```
\begin{tikzpicture}[scale=0.9,
    every node/.style={anchor=base, transform shape}]
  \draw[help lines] grid (13,1);
  \path foreach[count=\cnt] \ltr in {a,b,...,z} {
    (0.5*\cnt-0.25,0.4) node {\ltr} };
\end{tikzpicture}
```

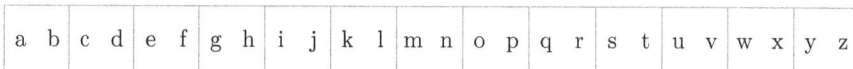

| a | b | c | d | e | f | g | h | i | j | k | l | m | n | o | p | q | r | s | t | u | v | w | x | y | z |

A single \foreach loop can have two loop variables, as shown in Example 4.9. Here we specify the text as \dir and the angle that should control its placement as \deg. The loop pairs are separated by a forward slash.

Example 4.9: Loop with paired variables

north

east

west

south

```
\begin{tikzpicture}[
    every node/.style={near end, auto}
]
\foreach \dir/\deg in
  {east/0,north/90,west/180,south/270}
  \draw[->] (0,0) -- (\deg:1.5cm) node {\dir};
\end{tikzpicture}
```

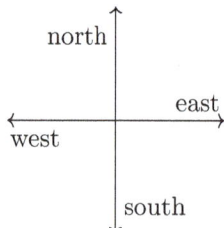

The `evaluate` option in a `foreach` loop is shown in Example 4.10, where we perform a calculation with the loop variable. The formatting of `evaluate` many not be intuitive, as what follows the equal sign is not the thing to be evaluated nor the variable being assigned, but rather the variable involved in the expression. The meaning of the expression is that \X, which is literally a string value such as "2^0", is evaluated as a math expression with the result stored in \Xeval. If we stopped there, the values would come out as decimals such as 1.0, so we go further and use `using` to give an expression involving \X and additional math operations, in this case `int` to convert the decimal to a whole number.

Example 4.10: Loop with variable evaluation

```
Preamble:\usetikzlibrary{math}
\begin{tikzpicture}[every node/.style={draw,align=center}]
  \pgfmathsetmacro{\nw}{width("$2^6=64$")}
  \foreach[count=\i, evaluate=\X as \Xeval using int(\X)] \X in
    {2^0,2^...,2^6} {
    \node[minimum width=\nw+1em] at (1.75*\i,0) {$\X=\Xeval$};
  }
\end{tikzpicture}
```

$2^0 = 1$	$2^1 = 2$	$2^2 = 4$	$2^3 = 8$	$2^4 = 16$	$2^5 = 32$	$2^6 = 64$

Example 4.11 is a more complicated `evaluate` example. In this case, the mathematical expressions *must* be contained in curly braces because they contain commas.

Example 4.11: Loop with multiple variable evaluation

```
Preamble:\usetikzlibrary{math}
\begin{tikzpicture}[every node/.style={anchor=base},xscale=0.9]
  \draw[help lines] grid (13,1);
  \foreach[count=\cnt,
    evaluate=\cnt as \xx using {mod(\cnt-1,13)+0.5},
    evaluate=\cnt as \yy using {0.4*(1-div(\cnt-1,13))+0.2}
  ] \ltr in {a,b,...,z} { \node at (\xx,\yy) {\ltr}; }
\end{tikzpicture}
```

a	b	c	d	e	f	g	h	i	j	k	l	m
n	o	p	q	r	s	t	u	v	w	x	y	z

4.6   Inline variables for math and coordinates

There are special inline variables that are defined in and scoped within a \path statement; see Reference 4.5. These inline variables enable using just the x-value or y-value of a coordinate as well as other calculations.

Reference 4.5: Inline variable assignment with `let ...in`

`\path ... let Assignment, Assignment,... in ...; % Needs calc library`

Assignment	Description
`\n`*Number*`=`*MathExpression*	Contain result of math expression
`\p`*Number*`=`*Coordinate*	Contain coordinate
`\x`*Number*`=`*XValue*	Contains x-value of coordinate `\p`*Number*
`\y`*Number*`=`*YValue*	Contains y-value of coordinate `\p`*Number*

Example 4.12 shows how to use path variables to set the `text width` using the distance between the x-values of two coordinates. The node (A) is created, and we want to put some text underneath in a second node such that the `text width` is 80% of the width of the first node. We set `\p1=(A.south west)` and `\p2=(A.south east)`, create an (invisible) path between them, and place a node midway and below. We use `\x1` and `\x2`, which are automatically defined by the assignment of `\p1` and `\p2`, to set the text width. Any distance calculated this way is in the unit of points, which is also the default unit of `text width`, and so does not need to be specified.

Example 4.12: Calculating a distance using path variables

$$f(x) = xe^x$$

Integrate this equation.

```
Preamble:\usetikzlibrary{calc,math}
\begin{tikzpicture}
  \node[draw,thick,inner sep=1em] (A) {$f(x)=xe^x$};
  \path let \p1=(A.south west),\p2=(A.south east) in
    (\p1) -- (\p2)
  node[text width=0.8*(\x2-\x1), blue, midway, below,
    font=\footnotesize\itshape,align=flush center]
  {Integrate this equation.};
\end{tikzpicture}
```

The `\n`*Number* expression can be used independently of coordinates. Example 4.13 re-creates Example 4.10 on the previous page using a path variable `\n1` for specifying the width of the nodes.

Example 4.13: Using inline path variables

```
Preamble:\usetikzlibrary{calc,math}
\begin{tikzpicture}[every node/.style={draw,align=center}]
  \path let \n1={width("$2^6=64$")} in foreach [count=\i,
    evaluate=\X as \Xeval using int(\X)] \X in {2^0,2^...,2^6} {
    (1.75*\i,0) node[minimum width=\n1+1em] {$\X=\Xeval$} };
\end{tikzpicture}
```

| $2^0 = 1$ | $2^1 = 2$ | $2^2 = 4$ | $2^3 = 8$ | $2^4 = 16$ | $2^5 = 32$ | $2^6 = 64$ |

4.7   Styles with arguments

We discussed declaring and using styles in Section 2.6. It is sometimes useful to have one or more optional arguments for a style. This is a capability of PGF, using its key management capabilities. For a single input argument to a defined style, simply use #1 to refer to that input, and use *StyleName*/.default=*Val* to set the default value of the input if desired.

Example 4.14 shows using an argument for the custom style topsy to specify the rotation of a node. For node A, no argument is given, so it uses the default of 180. For nodes B and C, the argument is given explicitly.

Example 4.14: Creating a new style (topsy) with argument and default value

```
\begin{tikzpicture}[
    topsy/.style={draw,circle,rotate=#1},
    topsy/.default=180
]
\path node[topsy] {A} ++(1,0) node[topsy=60] {B}
    ++(1,0) node[topsy=-45] {C};
\end{tikzpicture}
```

Example 4.15 is another example of a style with a single input. It declares the shorten style, which decreases *both* ends of a path by the same amount.

Example 4.15: Creating a new style (shorten) with argument

```
Preamble:\usetikzlibrary{positioning}
\tikzset{shorten/.style={shorten >=#1,shorten <=#1}}
\begin{tikzpicture}[every node/.style={circle,draw}]
  \node (A) {A};
  \node[above right=of A] (B) {B};
  \path[draw,shorten=2mm,<->] (A) -- (B);
\end{tikzpicture}
```

We are not limited to a single argument. Example 4.16 creates the matrix style with two arguments using .style n args, and it is also my favorite way to create a matrix. We first draw the 3×2 gray matrix on the left. Next, we add an equal sign to its right. We place the leftmost 3×2 blue matrix to the right of the equal sign but aligned with the top of the gray matrix. Note that we have set the anchor for the matrix to be north west. The diagonal 2×2 matrix is created in two steps. First, we create a 2×2 matrix but set fill=none *after* calling the matrix style. We also use xshift=2mm to move it over slightly. Then we explicitly draw the diagonal line. Next, we create the rightmost 2×2 blue matrix. Finally, we add labels above each matrix.

Example 4.16: Creating a new style (`matrix`) with two arguments

```
\tikzset{label/.style={above,font=\largeabove,font=\large},
  matrix/.style n args={2}{transform shape, thick, draw, fill=.!50,
    minimum height=#1 cm, minimum width=#2 cm,anchor=north west}}
\begin{tikzpicture}[scale=0.8]
  \node[gray,matrix={3}{2}] (A) {};
  \node[right,font=\large] at (A.east) (eq) {=};
  \node[blue,matrix={3}{2}] at (eq.east |- A.north) (U) {};
  \node[blue,matrix={2}{2},fill=none,xshift=2mm]
    at (U.north east) (S) {};
  \draw[blue,ultra thick] (S.north west) -- (S.south east);
  \node[blue,matrix={2}{2},xshift=2mm] at (S.north east) (V) {};
  \node[label] at (A.north) {$\mathbf{A}$};
  \node[label] at (U.north) {$\mathbf{U}$};
  \node[label] at (S.north) {$\mathbf{\Sigma}$};
  \node[label] at (V.north) {$\mathbf{V}$};
\end{tikzpicture}
```

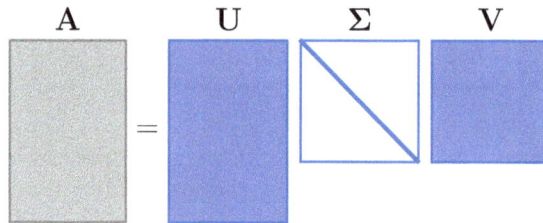

Part II: PGFPLOTS

Chapter 5

Enabling PGFPLOTS

PGFPLOTS is a package for plotting and is built on top of TikZ/PGF. PGFPLOTS enables users to create line, scatter, and bar plots. An advantage of PGFPLOTS is its tools for managing the myriad of elements involved in a plot, from reading data from files to setting up the axes, ticks, legends, etc. PGFPLOTS determines many plot elements automatically, such as setting axis limits based on the data, but the user can easily customize it as well. To use PGFPLOTS, you must include the pgfplots package in your preamble as shown in Reference 5.1.

Reference 5.1: Enabling PGFPLOTS (document preamble)

```
\usepackage{pgfplots}
```

Like TikZ, the functionality of PGFPLOTS can be extended with libraries. We will use the PGFPLOTS libraries listed in Reference 5.2. Examples that need these libraries will clearly indicate their use.

Reference 5.2: Optional PGFPLOTS libraries (document preamble)

```
\usepackage{pgfplotstable} % For transposing tables
\usepgfplotslibrary{groupplots} % For multiple aligned plots
\usepgfplotslibrary{fillbetween} % For filling between plots
```

The coverage of TikZ/PGF in Part I should make PGFPLOTS a bit easier to understand since PGFPLOTS relies on TikZ drawing and nodes. However, you can certainly jump right into PGFPLOTS without being an expert in TikZ.

The general setup for a PGFPLOTS graphic is shown in Reference 5.3. We need a tikzpicture to set up the canvas for the plot, but this generally needs no options. The plot itself is inside an axis environment. The axis options specify the size, title, axis labels, tick marks, and so on. Finally, one or more \addplot commands at the innermost level do the plotting, and plot options specify the color, line style, mark, etc. We focus on plotting data, so the coordinates for the \addplot come from a table, which reads data from a file.

Reference 5.3: General plot setup

```
\begin{tikzpicture}
  \begin{axis}[AxisOptions]
    \addplot[PlotOptions] table[TableOptions]{FileName};
  \end{axis}
\end{tikzpicture}
```

A way to specify PGFPLOTS options globally is via the \pgfplotsset command in Reference 5.4, which is analogous to \tikzset. We suggest some global defaults in Section 10.15.

Reference 5.4: \pgfplotsset

```
\pgfplotsset{...}
```

As mentioned earlier, our main focus is on plotting data from files. We give an overview of the extensive axis and plot options in Chapter 6, with pointers to detailed descriptions. In Chapter 7, we explain line, scatter, and bar plots for data files. In Chapter 8, we explain more about data files, including how to transform the data, provide inline tables, indicate missing data, etc. Basic refinements of plots are covered in Chapter 9, including changing fonts, number formatting, scaling of tick labels, tick marks and grids, legend options, aligning multiple axes using group plots, filling between two plots, and detaching a legend to be placed elsewhere. Advanced topics and nuances are covered in Chapter 10, including plotting mathematical expressions, combining TikZ and PGFPLOTS commands, and axis reference points.

Chapter 6

PGFPLOTS Overview

In this chapter, we give our pocket reference for PGFPLOTS, with pointers to detailed discussions in the remainder of the chapter. We list all the options covered for the axis in Reference 6.1 and addplot in Reference 6.2.

6.1 Axis overview

An axis environment is actually a special TikZ node. The axis command has four variants (axis, semilogxaxis, semilogyaxis, and loglogaxis) that determine which of the axes are linear or logarithmic. These axes have many dozens of options, so we cover only a subset in Reference 6.1. These options specify things such as the size of the axes, the title, the axis labels, the axis limits and ticks, the tick labels, and the legend. We group the keys into groups: axis format, axis limits and ticks, number formatting, tick and grid display, legend options, bar plots, and axis refinements. Additionally, there are styles that can be modified to impact the appearance of the plots. All of the options are discussed in detail in the sections that follow.

Many keys (or options/styles) have variants for both the x and y axes. We indicate that a key supports either x or y with a colored symbol ⓧy or xⓨ (green is used for keys in key-value pairs and orange is used for values or reserved keywords). The symbol in gray (xy) means that the axis specification can be omitted, in which case the option applies to both axes. As with TikZ, spaces matter in the options, though there is no particular rhyme or reason to whether or not a space follows x or y. For example, x dir is okay but xdir is not; in contrast, xmode is okay but x mode is not. Thus, we have variants of our special symbols that include an extra space character before or after the axis specifier; for example, scaled xy ticks means scaled ticks and scaled x ticks are valid, but not scaled xticks. And enlarge xy limits means enlargelimits and enlarge y limits are valid, but not enlargeylimits.

⚠ **Remember:** Spaces in the option keys matter!

Reference 6.1: Axis options overview

```
\begin{axis}[AxisOptions] PlotCommands \end{axis}
\begin{loglogaxis}[AxisOptions] PlotCommands \end{loglogaxis}
\begin{semilogxaxis}[AxisOptions] PlotCommands \end{semilogxaxis}
\begin{semilogyaxis}[AxisOptions] PlotCommands \end{semilogyaxis}
```

Axis format	Description
width=W, height=H	Width and height; see §7.2, §10.12
scale only axis	Axis size is *actually W × H*; see §7.2, §10.12
title={Title}	Title; see §7.2
title style={Style}	Append to every title style; see §9.1, §10.9
every axis title shift=S	Vertical shift of title, default 6 pt; see §10.9
xy label={Label}	Axis labels; see §7.2
xy label style={Style}	Append to axis label style; see §9.1, §10.10
font=Font	Font for plot elements; see §9.1
xy label near ticks	Put axis labels near tick labels; see §10.10

Axis limits and ticks	Description
xy min=Value, xy max=Value	Axis limits; see §7.4
enlarge xy limits=Option	Enlarge limits a little bit; see §7.4
xy tick distance=D	Distance between tick labels; see §7.4
xy tick={List}	Axis ticks; see §7.4
xy tickten={List}	Exponents of log axis ticks; see §7.4
xy ticklabels={List}	Labels for axis ticks; see §7.4

Number formatting	Description
xy ticklabel style={Style}	Append to every xy tick label; see §9.1
log ticks with fixed point	Write out numbers on log axis; see §9.2
scaled xy ticks=Option	Control number tick scaling; see §9.2

Tick/grid display	Description
ticks=Option	Show ticks: minor, major, both, or none; see §9.3
xy majorticks=TF	Show major ticks on axis; see §9.3
xy minorticks=TF	Show minor ticks on axis; see §9.3
minor xy tick num=N	Number of minor ticks between major ticks; see §9.3
major tick length=Len	Length of major tick; see §9.3
minor tick length=Len	Length of minor tick; see §9.3
tick style=Style	Append to every tick; see §9.3
grid=Option	Show grids; see §9.3
xy majorgrids=TF	Show major grid lines; see §9.3
xy minorgrids=TF	Show minor grid lines; see §9.3
grid style=Style	Append to every axis grid; see §9.3

Legend option | **Description**

legend entries={*List*} Legend entries; see §7.2, §9.4

legend pos=*Pos* Legend position; see §7.2, §9.4

legend columns=*N* Legend arrangement; see §9.4

legend cell align=*Align* Alignment of cells; see §9.4

legend style={*Style*} Append to every axis legend; see §9.1, §9.4

reverse legend Reverses the order of the legend entries; see §9.4

legend to name=*Name* Detach legend for, e.g., shared legend; see §10.7

Bar plot | **Description**

xybar Bar plot; see §7.6

xybar stacked Stacked bar plot; see §7.6

bar width=*W* Width of bars in bar plot; see §7.6

bar shift auto=*S* Space between multi-bar plots; see §7.6

tick align=*Align* Control alignment of ticks; §7.6

axis on top Redraw axis lines after drawing plots; see §7.6

Axis refinements | **Description**

xy dir=reverse Reverse the axis direction; see §10.2

xymode=*Option* Mode (linear or log); see §10.2

axis equal image Equalize axes units; see §10.2

symbolic xy coords={*List*} Specify order of symbolic coordinates; see §10.1

name=*Name* Name the axis node; see §10.5

at=(*C*) Coordinate for placement of the axes; see §10.5

anchor=*Anchor* Anchor for axis inside TikZ picture; see §10.5

cycle list={*List*} List of plot styles (not recommended); see §10.11

Style | **Description**

every axis For all axes such as width, height; see §7.2

every title style Title font, location, etc.; see §9.1, §10.9

every axis xy label Label font, locations, etc.; see §9.1, §10.10

every axis legend Legend font, location, etc.; see §9.1, §9.4.3

every xy tick label Tick label font, number formats, etc.; see §9.1

every tick Tick color, thickness; see §9.3

every outer xy axis line Axes, e.g., draw=none

every xy tick scale label Control formatting of tick scaling label

every axis grid Formatting of grid; see §6.1

log plot exponent style Style for exponents in log ticks; see §9.2

6.2 Plot overview

One we have an axis, we can add one or more plots. The \addplot command draws a plot, such as line, bar, or scatter. Its overview is provided in Reference 6.2. In subsequent chapters, we cover line plots (Section 7.3), scatter plots (Section 7.5), and bar plots (Section 7.6). Regardless of its type, the data source for the plot is most often a table that is read from a data file and used to specify the (x,y) coordinate pairs. Tables are covered briefly in Section 7.1 and in full detail in Chapter 8. A fill between plot fills in the space between two prior plots; see Section 9.6. Additionally, it is also possible to plot simple math expressions via expression; see Section 10.3.

For a given plot, we can specify the color, line style, mark type, mark size, and so on, with some options specific to the type of plot. As always, there are also styles that impact the plot.

Reference 6.2: Plot options

\addplot[*PlotOptions*] table[...]{*TableSource*}; (see §7.1 and Chapter 8)
\addplot[*PlotOptions*] fill between[of=*A* and *B*]; (see §9.6)
\addplot[*PlotOptions*] expression[...] {*MathExpression*}; (see §10.3)

Plot option	Description
Color	Color for all plot elements; see §7.3
LineStyle	Line style; see §7.3
mark=*Mark*	Specify mark shape; see §10.8
mark size=*Size*	Set mark size; see §10.8
mark color=*Color*	Set secondary mark color; see §10.8
only marks	No line, only marks; see §10.8
scatter	Enables marker modifications; see §7.5
scatter src=explicit symbolic	Use classes marker colors; see §7.5
scatter/classes={*Classes*}	Scatter plot classes; see §7.5
bar width=*W*	Bar plot bar width; see §7.6
fill=*Color*	Bar plot bar fill; see §7.6
forget plot	Do not reference plot in legend; see §9.4

Style	Description
every axis plot	Plot style, which can be set as a default
every axis post plot	After *PlotOptions* (useful for bar plots); see §7.6
every mark	Modify mark colors, rotation, etc.; see §10.8

Chapter 7

Plotting Basics

Our aim in this section is to get you quickly plotting. We first cover basic data file formats (Section 7.1) and then essential axis options such as axis size, title, and legend (Section 7.2). That lays the groundwork for line plots and plot markers in Section 7.3. We proceed to cover basic modifications of the axis limits and ticks (Section 7.4); scatter plots (Section 7.5); bar plots (Section 7.6); and user-defined plot styles (Section 7.7).

7.1 Basic data files

Data files should generally be formatted as follows.

- Data is formatted as a table, with an equal number of entries in every column.
- Columns are separated by spaces (default) or commas.
- Rows are separated by new lines.
- Comment lines begin with #.
- Entries containing one or more column separators (i.e., spaces) are encapsulated in curly braces.

An example data file looks something like what we see in `data.dat` (on this page). It has a comment on the first line, denoted by #. There are five columns and four rows of data, plus a row of column names.

Data File: data.dat

```
# Example data file
Id  A    B    C    D
1   6.5  5.2 10.8 12.3
2   5.0  8.5  9.4 11.3
3   3.3  7.1  6.3  3.5
4   2.2  4.1  7.9  3.9
```

We focus on plotting data from tables, so we use the `table` specifier to `\addplot` as shown in Reference 7.1. The key table options specify how columns are separated and which columns contain the x-values and y-values for the coordinates to be plotted.

We can also specify which column contains `meta` information to distinguish between different classes or categories of coordinates.

> **Reference 7.1: Basic `table` options for data files**
>
> `\addplot[...] table[TableOptions]{FileName};`
>
> | Table option | Description |
> |---|---|
> | `col sep=Colsep` | Options include `space` (default), `tab`, `comma`, `semicolon`, `&` |
> | `x=ColName` | Specify column for x-values; defaults to first column |
> | `y=ColName` | Specify column for y-values; defaults to second column |
> | `meta=ColName` | Specify column for class metadata for scatter plots |

In Example 7.1, we show our very first plot and illustrate the use of `table` to read data from `data.dat` (on the previous page). The x-values come, by default, from the first column. The table option `y=B` indicates that the y-values come from the column with header "B" (third column), overriding the default, which would have been the second column. Axis options are discussed in the next section (Section 7.2), and line plot options are discussed in Section 7.3.

> **Example 7.1: Very basic example of plotting data from a file**
>
> ```
> \begin{tikzpicture}
>   \begin{axis}
>     \addplot[blue] table[y=B]{data.dat};
>   \end{axis}
> \end{tikzpicture}
> ```
>
>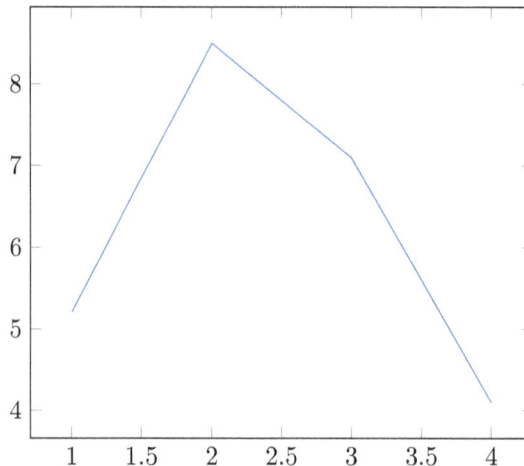

Many more details of handling data files are discussed in Chapter 8, including mathematical transformations, missing data, transposing data, etc.

7.2   Basic axis format and legend specification

Reference 7.2 gives the essential axis options, including type of axis (e.g., `loglogaxis`), axis size, title, labels for x and y axes, and legend entries and position.

Reference 7.2: Essential axis options

```
\begin{axis}[AxisOptions] PlotCommands \end{axis}
\begin{loglogaxis}[AxisOptions] PlotCommands \end{loglogaxis}
\begin{semilogxaxis}[AxisOptions] PlotCommands \end{semilogxaxis}
\begin{semilogyaxis}[AxisOptions] PlotCommands \end{semilogyaxis}
```

Axis option	Description
`width=W`	Width; default is 240 pt (8.4 cm)
`height=H`	Height; default is 207 pt (7.3 cm)
`scale only axis`	Axis size is *actually* $W \times H$; highly recommended!
`title=Title`	Graph title; default is none
`title style={Style}`	Append to `every title style`
`every axis title shift=S`	Vertical shift of title; default is 6 pt
`xy label=Label`	Label for axis; default is none
`xy label style={Style}`	Append to `every axis xy label`
`xy label near ticks`	Labels near tick labels; highly recommended!
`legend entries={List}`	Legend entries; default is no legend
`legend pos=Pos`	Legend position: `south west`, `south east`, `north west`, `north east` (default), or `outer north east`

Style	Description
`every axis`	Style for all axes, such as `width`, `height`, etc.
`every title style`	Modify title font, location, etc.; see §9.1, §10.9
`every axis xy label`	Modify label font, locations, etc.; see §9.1, §10.10

Together with `scale only axis`, the `width` and `height` specify the size of the axes themselves. The entire space required may be larger due to extra elements such as axis labels. Without `scale only axis`, the size of the axes are $(W - 45\,\text{pt}) \times (H - 45\,\text{pt})$. The distance 45 pt is approximately 1.5 cm. See Section 10.12 for further discussion of setting these parameters.

The `title`, `xlabel`, and `ylabel` specify the text to be used for the title, x-axis label, and y-axis label, respectively. These must be contained in curly braces if the text contains a comma. These are nodes, and their styles can be controlled via `title style`, `label style`, etc., which appends to the specified styles. It is highly recommended to specify both `xlabel near ticks` and `ylabel near ticks` to have the labels near the tick marks; see Section 10.10 for details.

The `legend entries` are matched with the plots in order; see Section 9.4 for further details and additional options for the positioning and formatting of the legend.

Remember: My recommended settings for `axis` options follow below, and Section 10.15 explains how to set these as defaults.

- `scale only axis` Sets `width` and `height` to the actual size of the axis rather than each being 45pt less than specified.

- `xlabel near ticks` Places the axis label directly underneath the x tick labels rather than a fixed distance of 15pt below the x-axis.

- `ylabel near ticks` Places the axis label directly to the left of the y tick labels rather than a fixed distance of 35pt to the left of the y-axis.

- `every axis title shift`=3pt Changes y-shift of title from the default of 6pt to 3pt.

Example 7.2 revisits Example 7.1 on page 56, adding axis settings. Here we use my recommended axis settings (as detailed above) and additionally specify the axis width and height, the title and axis labels, and the legend. The width and height refer to the size of the axes themselves, i.e., the box you see. The legend is in its default position, `north east`.

Example 7.2: Specifying axis options

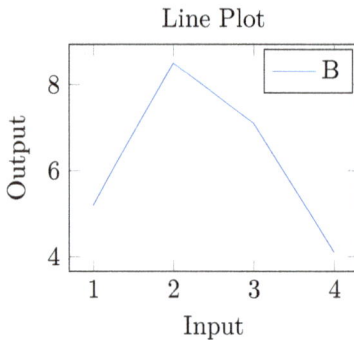

```
\begin{tikzpicture}
  \begin{axis}[scale only axis,
      xlabel near ticks,
      ylabel near ticks,
      every axis title shift=3pt,
      width=4cm, height=3cm,
      title=Line Plot,
      xlabel=Input, ylabel=Output,
      legend entries={B}]
    \addplot[blue] table[y=B]{data.dat};
  \end{axis}
\end{tikzpicture}
```

Without my recommended settings, the plot would appear as we shown in Example 7.3. The most striking difference is due to the omission of `scale only axis`. This causes the width and height to be 45 pt (1.5 cm) less than specified; see Section 10.12 for discussion. The placement of the x and especially the y labels differ; see Section 10.10. Finally, the title is 3 pt higher; see Section 10.9.

Example 7.3: Specifying axis options without my recommended defaults

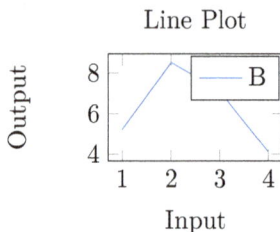

```
\begin{tikzpicture}
  \begin{axis}[
      width=4cm, height=3cm,
      title=Line Plot,
      xlabel=Input, ylabel=Output,
      legend entries={B}]
    \addplot[blue] table[y=B]{data.dat};
  \end{axis}
\end{tikzpicture}
```

7.3   Line plot

The default plot is a line plot. Reference 7.3 covers the basic options, including color, line style, and mark.

Reference 7.3: Basic line plot options

`\addplot`[*PlotOptions*] `table`[*TableOptions*]`{`*FileName*`};`

Plot option	Description
`Color`	Color for all plot elements; default is black
`LineStyle`	Line style like `thick` per options in Figs. 2.2a, §2.2c
`mark=`*Mark*	Mark symbol per Reference 7.4; default is `none` (no mark)
`mark size=`*Size*	Half the mark size; default is 2 pt
`mark color=`*Color*	Set secondary mark color; default is white
`only marks`	No line, only marks; see also scatter plot in §7.5

Style	Description
`every axis plot`	Plot style, which can be set as a default
`every mark`	Modify mark properties; see also §10.8

The *Color* and *LineStyle* set the overall style, including the style of the marks. The line is not drawn if `only marks` is specified, resulting in a scatter plot. (Scatter plots with multiples "classes" of data are discussed in Section 7.5.) The marks are set by `mark` using one of the choices in Reference 7.4 or `none` for no marks. The mark size is *twice* the `mark size`, which defaults to 2 pt. The `mark color` does *not* actually specify the mark color but rather the secondary mark color for two-color marks such as `halfdiamond*`.

Reference 7.4: List of available marks

We show each mark at the default `mark size=2pt` (2 pt is approximately half the size of the mark) and again with `mark size=5pt` in front of an orienting green box of size 10 pt × 10 pt centered at the point where the mark is placed. The last view allows us to see transparent and opaque regions.

> **?** **Oddity:** The size of a mark is *twice* the `mark size`; the `mark color` does *not*
> specify the (primary) mark color; and marks inherit the plot line style, even if it
> is not solid!

Example 7.4 shows a line plot. Options for `\addplot` specify the color and mark for the
plot; the * indicates a colored circle. The data comes from `data.dat` (on page 55). As
no `table` options are specified, by default the x-values come from the first column and
the y-values come from the second column. PGFPLOTS sets the axis limits automatically
to match the data.

Example 7.4: Simple line plot

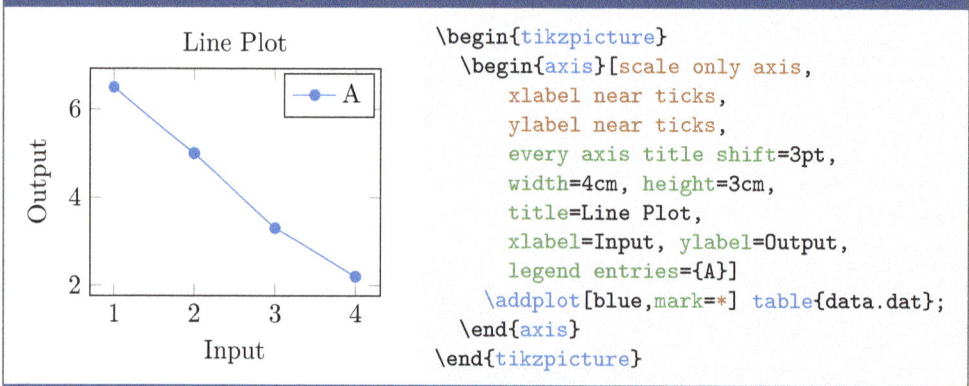

```
\begin{tikzpicture}
 \begin{axis}[scale only axis,
     xlabel near ticks,
     ylabel near ticks,
     every axis title shift=3pt,
     width=4cm, height=3cm,
     title=Line Plot,
     xlabel=Input, ylabel=Output,
     legend entries={A}]
    \addplot[blue,mark=*] table{data.dat};
 \end{axis}
\end{tikzpicture}
```

In terms of the plot options, I advocate directly specifying the plot style in the plot
options as we do in Example 7.4; otherwise, the style reverts to styles specified in
something called the `cycle list` (see Section 10.11).

As mentioned above, the marks as well as the lines are impacted by the *Color* and
LineStyle. This may have unintended consequences since specifying a dashed line
style will result in marks with dashed borders. The mark line style and (primary) color
can be changed by appending the `every mark` style. I recommend that the lines for
marks always be made solid by appending the `every mark` style to explicitly set the
line style to solid as shown in Reference 7.5 (this code is most effective in the preamble).

Reference 7.5: Ensuring solid mark lines, even if line style is not solid

```
\pgfplotsset{
  every axis plot/.append style={every mark/.append style={solid}}
}
```

Example 7.5 demonstrates adding multiple plots to the same set of axes. In this case,
both plots come from the same file, `data.dat` (on page 55), using the same x-values
(which defaults to the first column) and specifying different columns for the y-values.
We have plotted different colors and marks for each plot. As compared to the prior
example, PGFPLOTS automatically enlarges the y-axis so that all the data is visible.
Observe that we set the axis options using `\pgfplotsset`, including the special code
from Reference 7.5 to ensure the marks are drawn with solid borders.

Example 7.5: Multiple line plot with data from single file

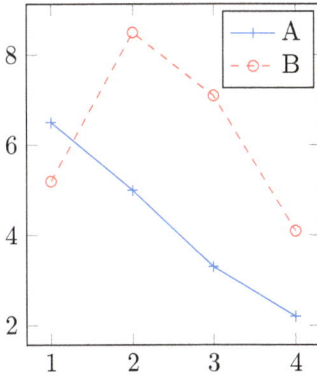

```
\pgfplotsset{scale only axis,
  width=4cm, height=4.5cm,
  every axis plot/.append style={
    every mark/.append style={solid}}}
\begin{tikzpicture}
  \begin{axis}[legend entries={A,B}]
    \addplot[blue,mark=+]
      table[y=A]{data.dat};
    \addplot[red,dashed,mark=o]
      table[y=B]{data.dat};
  \end{axis}
\end{tikzpicture}
```

Example 7.6 shows combining plots from two different data sources: `data.dat` (on page 55) is used to draw the blue line, and `points.csv` (on this page) is used for the red one. Since `points.csv` is comma-separated, we need `col sep=comma` in the table options for the second plot.

Data File: points.csv

```
x,      y,      class
0.458,  6.056,  a
0.757,  6.304,  a
0.988,  4.640,  a
2.924,  2.677,  b
4.426,  3.172,  b
4.746,  2.901,  b
```

Example 7.6: Multiple line plot with data from different files

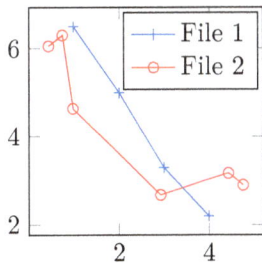

```
\begin{tikzpicture}
  \begin{axis}[scale only axis,
    width=3.2cm, height=3cm,
    legend entries={File 1, File 2}]
    \addplot[blue,mark=+] table{data.dat};
    \addplot[red,mark=o]
      table[col sep=comma]{points.csv};
  \end{axis}
\end{tikzpicture}
```

The final demonstration, in Example 7.7, is a log-log plot using `compress.csv` (on the next page), which is comma-separated. In this case, we set comma-separated as the default using `\pgfplotsset`. We specify the plot line style to be `thick`, which also impacts the line style of the marks.

Data File: compress.csv

```
Error, hcci_627, tjlr_16,  sp_50
1e-06, 1.89e+00, 1.70e+00, 4.54e+00
1e-05, 3.50e+00, 1.92e+00, 1.59e+01
1e-04, 8.15e+00, 2.57e+00, 5.51e+01
1e-03, 2.54e+01, 6.83e+00, 2.31e+02
1e-02, 1.38e+02, 3.69e+01, 5.58e+03
```

Example 7.7: Log-log plot

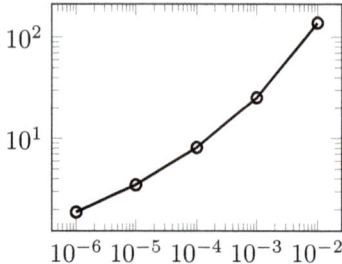

```
\pgfplotsset{table/col sep=comma}
\begin{tikzpicture}
  \begin{loglogaxis}[scale only axis,
    width=4cm, height=3cm]
    \addplot[thick, mark=o]
    table{compress.csv};
  \end{loglogaxis}
\end{tikzpicture}
```

7.4   Setting the axis limits and ticks

The options for setting axis limits and ticks are shown in Reference 7.6. By default, PGFPLOTS selects the axis limits and ticks automatically based on the data. Rather than setting the limits to exactly the min and max values in the data, they are expanded by 10%, which is controlled by enlarge $_{xy}$ limits. The ticks can be specified manually via $_{xy}$ tick; the option data pulls the ticks from the data, and the option \empty (with a backslash) indicates that there should be no ticks. Another nice option is $_{xy}$ tick distance, which specifies only the distance between ticks and generally does what one would desire. For log plots, $_{xy}$ tickten enables specifying only the exponents.

Oddity: The key for specification of the ticks is singular (xtick) whereas the key for specification of tick labels is plural (xticklabels).

By default, xmin and xmax for a linear axis are set as follows: assume $x_{\min}$ and $x_{\max}$ are the minimum and maximum x-values in the plots. Then

$$\text{xmin} = x_{\min} - 0.10 \, (x_{\max} - x_{\min}) \quad \text{and} \quad \text{xmax} = x_{\max} + 0.10 \, (x_{\max} - x_{\min}).$$

If xmin is set manually, then that limit is not enlarged unless enlargelimits=true explicitly. The enlargement effect is multiplicative for logarithmic axes; in other words, it is scaled in such a way as to provide a visual 10% buffer on the end of each axis. The proportion of enlargement can be specified via enlargelimits=*Value* (both axes) or for the individual axes. The default *Value* is 0.1. Automatic axis enlargement can be disabled via enlargelimits=false.

Reference 7.6: Axis options to set axis limits and ticks

`\begin{axis}[`*AxisOptions*`]` *PlotCommands* `\end{axis}`

Axis option	Description
`xy`min=*Value*	Lowest axis value
`xy`max=*Value*	Highest axis value
enlarge `xy` limits=*Option*	Modifications of axis limits; default `auto`
`xy`tick distance=*D*	Distance between ticks
`xy`tick={*T1,T2,...,TN*}	Specific ticks, `data`, or `\empty` for none
`xy`tickten={*E1,E2,...,EN*}	Base-10 exponents for ticks on log scale
`xy`ticklabels={*L1,L2,...,LN*}	Labels for specified ticks

Enlarge option	Description
`auto`	Enlarge limits except those specified via `xy`min or `xy`max
`true`	Enlarge limits, even those manually specified, by 10%
`false`	Do not enlarge the limits
Value	Enlarge all limits by the specified proportion

Example 7.8 uses `data.dat` (on page 55) to demonstrate `enlargelimits=false`, which causes the limits to be exactly the minimum and maximum of the data, so the axes are tight.

Example 7.8: Disabling `enlargelimits`

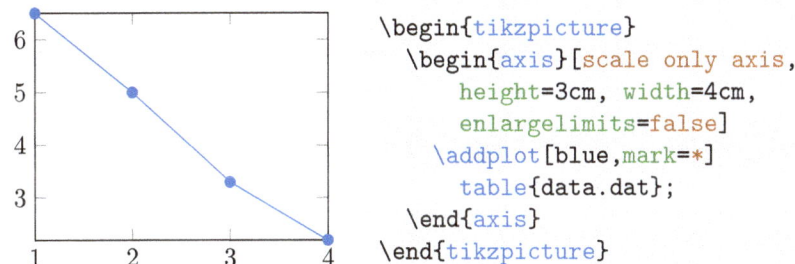

```
\begin{tikzpicture}
  \begin{axis}[scale only axis,
      height=3cm, width=4cm,
      enlargelimits=false]
    \addplot[blue,mark=*]
      table{data.dat};
  \end{axis}
\end{tikzpicture}
```

The ticks themselves are automatically determined unless an option such as `xtick` is specified. The `xtick distance=`*D* option sets the distance between subsequent ticks, but it is still the onus of PGFPLOTS to pick the starting point, which it does well. If that is insufficient, then the ticks can be specified manually. You specify sequences of tick marks using pattern-based syntax; for instance, `ytick={1,2,...,10}` is valid. The specification of ticks does not affect the axis limits, and only ticks within the limits are actually shown. Using `xtick=data` pulls the ticks from the data, using all the unique *x*-values. For specifying logarithmic ticks, the shorthand `ytickten={0,1,...,10}` is equivalent to `ytick={10^0,10^1,...,10^(10)}`. Finally, it is possible to specify the labels for the ticks using `xticklabels`. Number formatting and scaling of numeric tick labels is discussed in Section 9.2. The style of tick labels (font, rotation, etc.) is discussed in Section 9.1.

In Example 7.9, we manually specify `xtick={0,1,...,5}`; observe that 0 is below `xmin` (which was set automatically) and so is not shown. We specify `ytick distance=1`, and the resulting ticks on the y-axis become (2,3,...,6). This example uses `data.dat` (on page 55) and `points.csv` (on page 61).

Example 7.9: Specifying ticks

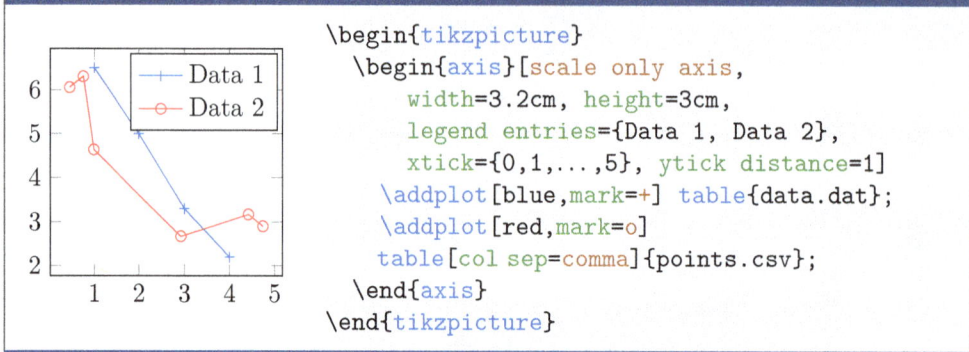

```
\begin{tikzpicture}
  \begin{axis}[scale only axis,
      width=3.2cm, height=3cm,
      legend entries={Data 1, Data 2},
      xtick={0,1,...,5}, ytick distance=1]
    \addplot[blue,mark=+] table{data.dat};
    \addplot[red,mark=o]
      table[col sep=comma]{points.csv};
  \end{axis}
\end{tikzpicture}
```

Example 7.10 is an example of manually specifying the tick labels using `xtick = {1,2,3,4}` and `xticklabels = {One,Two,Three,Four}`. This example uses `data.dat` (on page 55). Logically, the number of ticks needs to match the number of tick labels; any missing tick labels are left blank and extra ones are ignored.

Example 7.10: Explicit tick labels

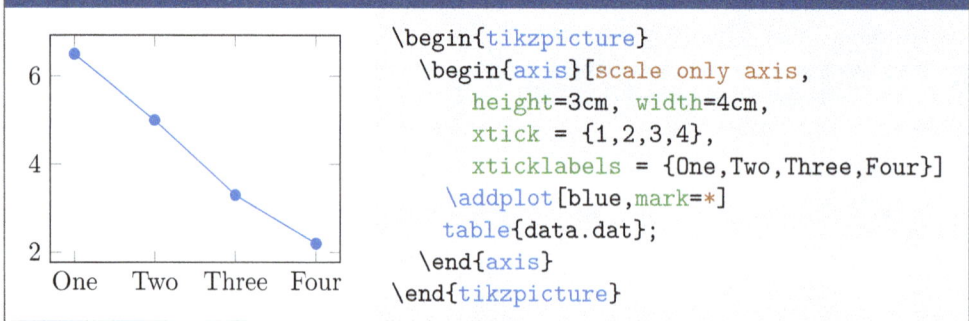

```
\begin{tikzpicture}
  \begin{axis}[scale only axis,
      height=3cm, width=4cm,
      xtick = {1,2,3,4},
      xticklabels = {One,Two,Three,Four}]
    \addplot[blue,mark=*]
      table{data.dat};
  \end{axis}
\end{tikzpicture}
```

7.5   Scatter plot

Scatter plots are often useful when comparing different sets of points plotted with different styles. We refer to the different sets as *classes*. If the set of points for each class is in its own file, then a scatter plot can be accomplished using the same structure as for line plots but with the `only marks` plot option. Example 7.11 is such an example. We specify `every axis plot/.style={only marks}` in the axis options. As a result, the first plot creates only blue circles, and the second produces only red triangles. This is essentially a line plot without the lines. This example uses `data.dat` (on page 55) and `points.csv` (on page 61).

Example 7.11: Scatter plot with one file per class

Scatter Plot

```
\begin{tikzpicture}
  \begin{axis}[scale only axis,
      width=3cm, height=3cm,
      title=Scatter Plot,
      legend entries={File 1, File 2},
      every axis plot/.style={only marks}]
    \addplot[blue,mark=*] table{data.dat};
    \addplot[red,mark=triangle*,mark size=3pt]
      table[col sep=comma]{points.csv};
  \end{axis}
\end{tikzpicture}
```

However, a common situation is to have all the data in one file, with the class indicated by a metadata column. For instance, the `class` column of `points.csv` (on page 61) indicates the class of each point. This can be handled as demonstrated in Reference 7.7. The option `scatter` indicates that the marks should be styled according to something associated with each point, such as the metadata. The option `scatter src=explicit symbolic` indicates that the scheme is based on a symbolic (e.g., text-based) rather than a numerical value (see the PGFPLOTS manual for styling based on numerical values). The `scatter/classes` option is specific to setting `scatter src=explicit symbolic` and defines the mark style for each class.

Reference 7.7: Scatter plot with classes

```
\addplot[only marks, scatter, scatter src=explicit symbolic,
  scatter/classes={
    Class1={Style1}, ... , ClassN={StyleN} % No final comma
  }]
  table[meta=ColName,...]{FileName};
```

⚠ **Remember:** In `scatter/classes`, the last class definition has no trailing comma!

The `meta` key in the table options specifies the column name for the column containing the class labels. The labels are handled as symbolic, so they can be numbers, letters, or words (case sensitive). The arguments to scatter classes are of the form `ClassK={StyleK}` where `ClassK` is one of the class labels and `StyleK` is the style for that class, usually a color and a mark. There should be no trailing comma after the final class specification. The `only marks`, `scatter`, `scatter src=explicit symbolic` options can alternatively be put into the `axis` options, but not `scatter/classes`.

In Example 7.12, we show a scatter plot from the file `points.csv` (on page 61). The `meta=class` option for `table` specifies the column containing the symbolic class labels. The `scatter/classes` option for `\addplot` gives the style for each class: blue circles for class `a` and red squares for class `b`. The classes can be listed in any order. The first item in `legend entries` will be associated with the first class listed, the second with the second class listed, and so on.

Example 7.12: Scatter plot

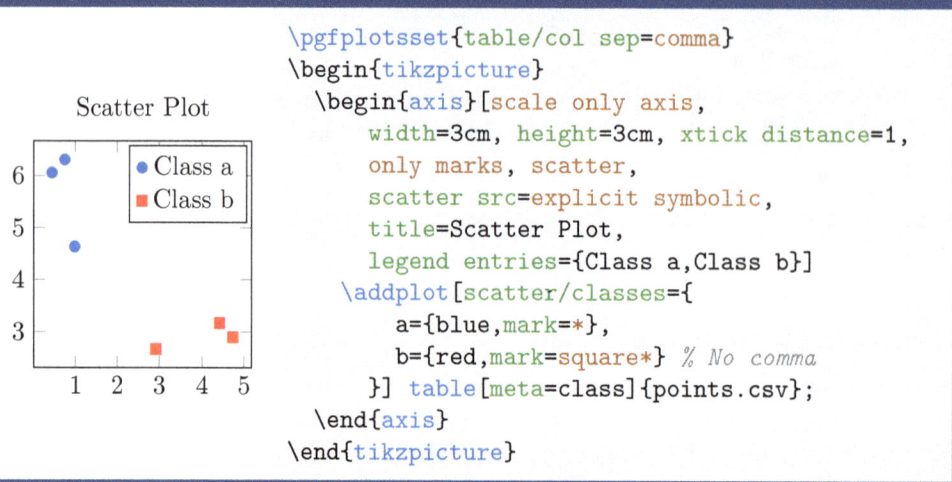

Scatter Plot

```
\pgfplotsset{table/col sep=comma}
\begin{tikzpicture}
  \begin{axis}[scale only axis,
      width=3cm, height=3cm, xtick distance=1,
      only marks, scatter,
      scatter src=explicit symbolic,
      title=Scatter Plot,
      legend entries={Class a,Class b}]
    \addplot[scatter/classes={
        a={blue,mark=*},
        b={red,mark=square*} % No comma
      }] table[meta=class]{points.csv};
  \end{axis}
\end{tikzpicture}
```

7.6 Bar plot

Bar plots can be horizontal or vertical, and they can be side by side or stacked. The options are summarized in Reference 7.8.

Reference 7.8: Bar plot options

`\begin{axis}[AxisOptions] PlotCommands \end{axis}`

Axis option	Description
xbar	Horizontal bar plot
ybar	Vertical bar plot
xybar stacked	Stacked bar plot
bar width=W	Bar width; default is 10 pt
bar shift auto=S	Bar shift for multiple bar plots; default is 2 pt
tick align=Align	Options are inside, outside, center
axis on top	Redraw axis lines after drawing bars

Plot option	Description
Color	Color for plot elements; default is black
draw=Color	Outline color for bar plot; default is black
fill=Color	Fill color; default is no fill

Style	Description
every axis plot post	Applied after PlotOptions; useful for bar plots

In contrast to scatter plots, bar plots are indicated as an axis option. The bars can be vertical (ybar) or horizontal (xbar). For multiple plots, the bars can be side-by-side (ybar) or stacked (ybar stacked). The bar width determines the width of all bars, and bar shift auto sets the spacing between bars in multi-bar plots. PGFPLOTS automatically sets the alignment of the ticks to be *outside* the axes in the direction of the bars. To change this back to normal, use `tick align=inside`.

> ⚠ **Remember:** Bar plots should be specified at the axis level.

The default bar is just an outline in the color specified in the `\addplot` options, with black as the default. The option `draw=Color` can be used to set the outline color. The option `fill` by itself sets the fill color to the plot's specified color as well. Alternatively, a different color can be set, e.g., `fill=.!50` sets the color to 50% of the designated color.

The `every axis plot post` style is applied after the plot options and is especially useful to set the fill based on the plot options. For example, `every axis plot post/.append style={fill,draw=black}` specifies that all bars are filled with the default color and outlined in black.

> **?₂?** **Oddity:** Sometimes filled bars color over the axis line; this can be remedied via the axis option `axis on top`, which redraws the axes after the bars have been drawn.

A bar plot is shown in Example 7.13 and uses `data.dat` (on page 55). We specify that it is a bar plot by adding `ybar` to the axis options. Because of the width of the bars, the axis limits can be somewhat tight, so we set `enlarge x limits=0.2`. The bars are unfilled because we did not specify `fill`.

Example 7.13: Basic bar plot

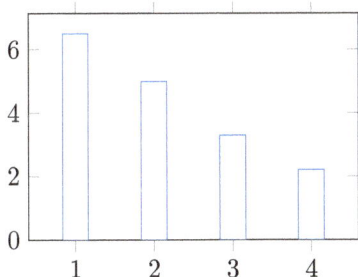

```
\begin{tikzpicture}
  \begin{axis}[scale only axis,
      width=4.5cm, height=3cm,
      ybar, ymin=0,
      enlarge x limits=0.2
    ]
    \addplot[blue] table{data.dat};
  \end{axis}
\end{tikzpicture}
```

Example 7.14 shows a bar plot with multiple bars using `runtime.csv` (on the next page). This plot has a logarithmic y-axis, indicated by the environment `semilogyaxis`. The `ybar` key specifies that this plot uses vertical bars. For multibar plots, the bars are shifted automatically so that they are centered as a group around the x-value. We specify `xtick=data` to indicate that the x ticks should be all x-values in the dataset, which is just $\{1, 2, 3, 4\}$ in this case. To improve spacing at either end of the plot, we increase the x limits using `enlarge x limits=0.2`. We provide labels for both axes and legend entries. Note that the legend entries are automatically formatted to indicate that these are bar plots. We position the legend using `legend pos=outer north east`. We set `every axis plot post/.append style={fill,draw=black}` to set the fill color to the designated bar color and draw the outline in black. We add the key `axis on top` because otherwise the bars overlap the x-axis.

Data File: runtime.csv

```
Num, Ltr, M1, M2, M3, M4
 1, A, 53.2, 617, 12.6, 94.1
 2, B, 104, 694, 20.6, 141
 3, C, 158, 6.87E+03, 47.9, 3.50E+03
 4, D, 11419, 1.07E+04, 2585, 5.51E+03
```

Example 7.14: Multi-bar plot

```
\pgfplotsset{scale only axis, xlabel near ticks, ylabel near ticks,
    table/col sep=comma}
\begin{tikzpicture}
  \begin{semilogyaxis}[width=3.4in, height=1.25in,
      ybar, xtick=data, enlarge x limits=0.2,
      ylabel=Run time (sec), xlabel=Dataset,
      legend entries={Method 1, Method 2, Method 3, Method 4},
      legend pos=outer north east,
      every axis plot post/.append style={draw=black, fill},
      axis on top]
    \addplot[blue] table[y=M1] {runtime.csv};
    \addplot[orange] table[y=M2] {runtime.csv};
    \addplot[gray] table[y=M3] {runtime.csv};
    \addplot[yellow] table[y=M4] {runtime.csv};
  \end{semilogyaxis}
\end{tikzpicture}
```

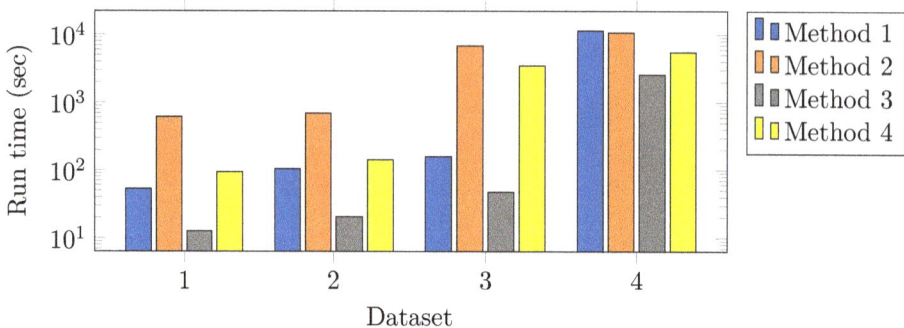

An example stacked bar plot, oriented horizontally, follows in Example 7.15. We pass
xbar stacked as an axis option. This example uses data.dat (on page 55). Because
our bars are being drawn horizontally, we need to swap the x- and y-values in the table
options, e.g., table[y=Id,x=A]{data.dat}.

⚠ **Remember:** Swap the x and y inputs to table when switching from a ybar
to an xbar plot.

Example 7.15: Stacked bar plot, oriented in x direction

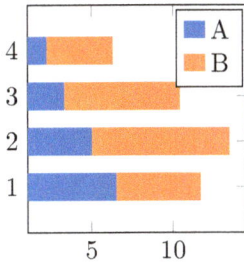

```
\begin{tikzpicture}
  \begin{axis}[scale only axis,
      width=3cm, height=3cm, xbar stacked,
      ymin=0, ymax=5, ytick={1,2,3,4},
      legend entries={A,B}]
    \addplot[blue,fill] table[y=Id,x=A]{data.dat};
    \addplot[orange,fill] table[y=Id,x=B]{data.dat};
  \end{axis}
\end{tikzpicture}
```

As mentioned above, the xbar or ybar option should be included as an *axis* option rather than a *plot* option. This is because it impacts not only the plot itself, but some other features such as the legend graphic and the automatic horizontal shifting of bars when there are multiple plots. It is possible to specify the legend graphic style and horizontal shift manually; see ⚏bar legend style and bar shift in the PGFPLOTS manual. See also Section 10.13 on mixing bar and line plots.

7.7 Defining and using plot styles

As with TikZ, it is possible to declare your own style keys; see Sections 2.6 and 4.7 for further details on TikZ styles. This can be useful for plot styles that are used repeatedly. We give an example of defining four plot styles in Example 7.16. We use \pgfplotsset to declare the styles a, b, c, and d. This declaration can be made in the preamble, which would make the styles accessible for all plots. This example uses data.dat (on page 55).

Example 7.16: Defining plot styles

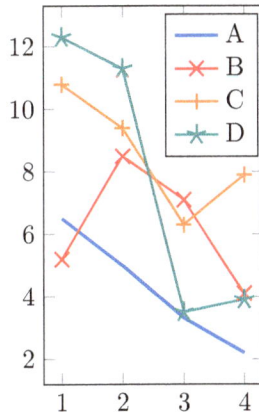

```
\pgfplotsset{
  scale only axis, width=3cm, height=5cm,
  a/.style={blue,very thick},
  b/.style={red,mark=x,thick,mark size=4pt},
  c/.style={orange,mark=+,thick,
    mark size=3pt},
  d/.style={teal,mark=star,thick,
    mark size=4pt}
}
\begin{tikzpicture}
  \begin{axis}[legend entries={A,B,C,D}]
    \addplot[a] table[y=A]{data.dat};
    \addplot[b] table[y=B]{data.dat};
    \addplot[c] table[y=C]{data.dat};
    \addplot[d] table[y=D]{data.dat};
  \end{axis}
\end{tikzpicture}
```

Chapter 8

Tables and Data Files

We discussed the basic `table` options and data file formats in Section 7.1. Before we discuss advanced plotting options in Chapter 9, we elaborate further on tables and data files, including transforming the input data, declaring tables inline, and handling missing data. We provide an overview of the options for working with tables in Reference 8.1.

Within an `\addplot` command, the `table` directive reads a table from an external file or inline data and specifies how to use the data in the table. This can be separated into reading the table via `\pgfplotstableread` (requiring the `pgfplotstable` LaTeX package) and then referencing the saved table in a `table` directive. Because these tasks can be separated, we also separate the options into two categories: the read options specify things such as the column separator or search path, and the table options specify things such as the x- and y-values.

Comment lines inside a data file start with a # symbol. Alternatively, you can use the option `skip first n` to skip a specified number of header lines. Data must be formatted as a table, with rows and columns. Column headers are optional; if used, every column header must be unique. The default column separator is one or more spaces (`col sep=space`), and the default row separator is a new line (`row sep=newline`). Blank or multi-word column headers and table cells must be enclosed within curly braces {}; see `res.dat` (on page 73) for an example. The value `nan` (not a number) is recommended to indicate a blank entry. You can make comma-separated format the default for tables using `\pgfplotsset{table/col sep=comma}` in the header.

White space can confuse PGFPLOTS, primarily when reading column headers or non-numeric values. Leading and trailing spaces can be eliminated by using `trim cells=true` to strip off any white space before or after an entry.

> **Remember:** If there are problems with referencing column names, try adding `trim cells=true` to the table options to gobble any leading or trailing white space.

Reference 8.1: Advanced data table options

```
Preamble:\usepackage{pgfplotstable} % For pgfplotstableread
\addplot[...] table[ReadOptions,TableOptions]{FilenameOrInline};
\pgfplotstableread[ReadOptions]{FilenameOrInline}{\SavedTable}
\addplot[...] table[TableOptions]{\SavedTable};
```

Read option	Description
col sep=*Colsep*	space (default), tab, comma, semicolon, &
row sep=*Rowsep*	newline or \\; last line must have row separator
skip first n=*N*	Skip first N lines (file only)
search path={*Path*}	Search path for files, comma separated
trim cells=true	Trim leading and trailing spaces in table entries

Table option	Description
x=*ColName*	Column name for x-values; default is first column
y=*ColName*	Column name for y-values; default is second column
meta=*ColName*	Column name for meta values; not defined by default
xy index=*N*	Column index (starting at zero) for value
meta index=*N*	Column index (starting at zero) for meta values
xy expr=*MathExpr*	Math expression for value
meta expr=*MathExpr*	Math expression for meta value

Math expression	Description
\thisrow{*ColName*}	Value in column *ColName*
\thisrowno{*ColIndex*}	Value in column index *ColIndex*, starting at zero
\coordindex	Row index (zero-based), ignoring header rows

Reference 8.2: Set data file search path

```
\pgfplotsset{table/search path={Dir1,Dir2,...}}
```

By default, the search path for data files is the current directory. Reference 8.2 shows how to modify this by using the search path option, where *DirX* represents an absolute or relative search path. Specify . (dot) to include the current directory. For example, search path={.,../data} includes the compilation directory and the directory data in the parent directory.

Every row in a table must have the same number of columns. If this is not the case, it is generally best to split the data into separate files. Another option is to fill out the shorter rows with nan. You obtain the x- and y-values for a plot by specifying table columns. If your tables are row-oriented data, see the discussion on transposing tables in Section 8.4.

As already demonstrated, the table options allow the user to specify the x- and y-values for a plot as well as an optional meta value that is useful for scatter plots. We have already seen column headers specified by name in the previous chapter. A column

Data File: res.dat

```
# Del Valle Reservoir (California) Storage
# taf:thousand acre feet, af:acre feet
year {min taf} {max taf} {min af} {max af}
2018 24.966 40.006 24966 40006
2019 24.975 40.876 24975 40876
2020 25.17 38.626 25170 38626
2021 28.926 40.633 28926 40633
2022 34.393 42.874 34393 42874
```

Example 8.1: Data transformation

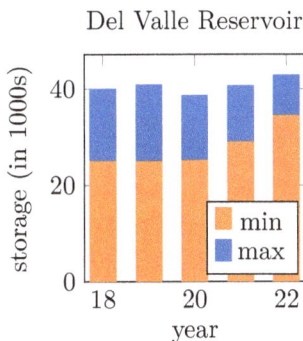

```
\pgfplotsset{scale only axis,
  width=3cm, height=3cm,
  xlabel near ticks, ylabel near ticks}
\begin{tikzpicture}
  \begin{axis}[ymin=0, ybar stacked,
    title=Del Valle Reservoir,
    ylabel=storage (in 1000s), xlabel=year,
    legend entries={min,max},
    legend pos=south east ]
  \addplot[orange,fill] table[
    x expr=\coordindex+18,
    y expr=\thisrowno{3}/1000]{res.dat};
  \addplot[blue,fill] table[
    x expr=\thisrow{year}-2000,
    y expr=(\thisrow{max af}-
      \thisrow{min af})/1000]{res.dat};
  \end{axis}
\end{tikzpicture}
```

can alternatively be specified by its index; for instance, x index=0 indicates that the x-values come from the first column. Specifications such as x expr provide access to rudimentary tools to process the table data via math expressions; see Section 8.1 for further details.

8.1 Data transformations

Math expressions can be used to compute derived values from the input data by using x expr, y expr, and meta expr as table options. Valid math expressions are listed in Section 4.4. The function \coordindex returns the zero-based index of the current row. PGFPLOTS provides several functions that can be invoked within these expressions, allowing you to reference the row's index or the value(s) of one or more columns. For example, the function \thisrow{ColName} returns the value specified in the named column for the current row. The command \thisrowno{N} returns the value specified in column N for the current row, with indexing starting at 0 (zero). Contents of multiple columns can be combined. Example 8.1 shows how to use math expressions to derive new values from the input data in res.dat (on the current page).

> ⚠ **Remember:** Row and column numbering starts at zero.

In the first `table` command, we use `x expr=\coordindex`+18 to calculate the year without the leading "20". The `\coordindex` gives the row number, starting at zero and ignoring the header row. We set the y-value to the value of column index 3 (fourth column, labeled `min af`) divided by 1000. This converts the units of the y-values from acre-feet to thousands of acre-feet. In the second plot command, we calculate the x-value a different (and arguably much more robust) way, using `\thisrow {year}`-2000. (I personally find the reference to the "row" in `\thisrowno` to be slightly confusing because it is referencing a column, but the intent is to take the contents of the specified column from *this row*.) Finally, we are doing a stacked plot and really just want to stack the difference between the two y-values as our second "stacked" contribution. We do this by taking the difference of two columns via `y expr=(\thisrow{max af}-\thisrow{min af})/1000`. We have also divided by 1,000 to convert the units.

8.2   Inline table declaration

It is possible to declare a table inline by just listing the entries where you would normally put the filename. With `\pgfplotstableread`, the table is saved as a macro for use in subsequent `table` commands. There can be no comments in an inline table, and the last row must be followed by the row separator. Headers are optional, as with tables in a file.

> ⚠ **Remember:** Inline tables cannot have comments and must terminate with a row separator.

In Example 8.2, we declare the inline table using LaTeX table format, specifying table options that set the column separators to ampersands and the row separators to \\. The table has two columns and four rows. We reference the table using `\mydata` in the subsequent `table` commands. We use the row index as the x-value via `x expr=\coordindex`. We use the column index to specify the y-value via, e.g., `y index=0`.

Example 8.2: Inline table declaration

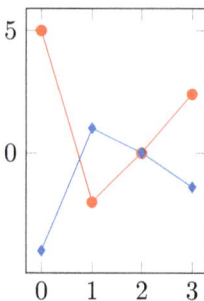

```
Preamble: \usepackage{pgfplotstable}
\pgfplotstableread[col sep=&,row sep=\\]%
{ 5 & -4 \\ -2 & 1 \\ 0 & 0 \\ 2.4 & -1.4 \\ }{\mydata}
\begin{tikzpicture}
  \begin{axis}[scale only axis, width=2.5cm,
      height=3.5cm]
    \addplot[red,mark=*]
      table[x expr=\coordindex, y index=0]{\mydata};
    \addplot[blue,mark=diamond*]
      table[x expr=\coordindex, y index=1]{\mydata};
  \end{axis}
\end{tikzpicture}
```

8.3   Missing data

In some cases, data points may be missing or unbounded. This can be handled by leaving that data blank or writing in nan, inf, or -inf. By default, unbounded co-ordinates are discarded. (This can be changed via the key unbounded coord; see the PGFPLOTS manual for details.)

In Example 8.3, we declare an inline data table using \pgfplotstableread. The default row separator is a new line, which we use here. It is critical to have a new line after the last row. There is a missing entry in the second row (not counting the header), indicated by nan. The \addplot command simply ignores this missing point.

Example 8.3: Missing data

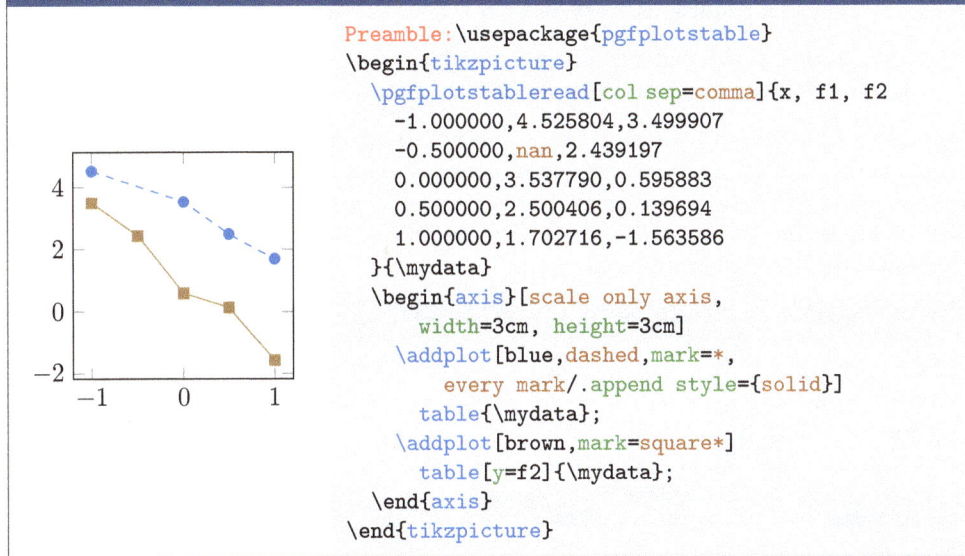

```
Preamble:\usepackage{pgfplotstable}
\begin{tikzpicture}
  \pgfplotstableread[col sep=comma]{x, f1, f2
    -1.000000,4.525804,3.499907
    -0.500000,nan,2.439197
    0.000000,3.537790,0.595883
    0.500000,2.500406,0.139694
    1.000000,1.702716,-1.563586
  }{\mydata}
  \begin{axis}[scale only axis,
      width=3cm, height=3cm]
    \addplot[blue,dashed,mark=*,
        every mark/.append style={solid}]
      table{\mydata};
    \addplot[brown,mark=square*]
      table[y=f2]{\mydata};
  \end{axis}
\end{tikzpicture}
```

In plotting the first line, we had to add every mark/.append style={solid} to the options so that the lines around the marks were not also dashed; see Reference 7.5 for the setting to avoid this problem.

LaTeX will print out a compilation message for missing data points. For instance, Example 8.3 includes the following output:

```
NOTE: coordinate (-0.250000,nan) has been dropped because it is unbounded
(in y). (see also unbounded coords=jump).
```

This can be suppressed by setting \pgfplotsset{filter discard warning=false}.

8.4   Transposing tables

As mentioned previously, PGFPLOTS expects data to be column-oriented in that columns correspond to x- and y-values. But sometimes data is given to you in a row-oriented format. Generally, it is recommended to externally preprocess the file to convert it from row-oriented to column-oriented. Nevertheless, there are some limited tools for transposing data tables available in the pgfplotstable package. Reference 8.3 details how to transpose a row-oriented table.

Reference 8.3: Transpose table

Preamble:\usepackage{pgfplotstable}
\pgfplotstabletranspose[colnames from=*Pivot*,
  input colnames to=*Pivot*]{\ByColTbl}{\ByRowTbl}

The output table (\ByColTbl) is the *first* argument of \pgfplotstabletranspose, and the input table (\ByRowTbl) is the second argument. We do not cover all possible ways of doing a table transpose but instead consider only one case. We let *Pivot* denote the contents of the (0,0) entry. In the input file, *Pivot* is the column header of the column that contains the future column names. In the output table, this same entry becomes the column header of the first column. For this transpose operation to work, the following conditions must be met.

- Every value in the first row of the input file must be unique!
- Every row must have the same number of entries.

Example 8.4 provides a demonstration using **rowdata.dat** (on this page), followed by a listing of the resulting table stored in \bycol. We can transpose this table using the incantation in Example 8.4. Spaces can cause problems, so we use trim cells=true when reading the row-formatted data. We use the special \pgfplotstabletypeset command to display the table contents.

Data File: rowdata.dat

```
idx        1   2   3   4   5
{data 1} 0.4 0.3 0.1 0.0 0.2
{data 2} 0.1 0.2 0.3 0.4 0.3
```

Example 8.4: Transposing table

Preamble:\usepackage{pgfplotstable}
\pgfplotstableread[trim cells=true]{rowdata.dat}{\byrow}
\pgfplotstabletranspose[colnames from=idx,input colnames to=idx]
  {\bycol}{\byrow}
\pgfplotstabletypeset[string type,font=\ttfamily]\bycol

```
idx    data 1   data 2
 1      0.4      0.1
 2      0.3      0.2
 3      0.1      0.3
 4      0.0      0.4
 5      0.2      0.3
```

Chapter 9

Refinements

There are seemingly infinite refinements one can make with PGFPLOTS. We focus here on those refinements that we think will generally be most useful in plotting various datasets from experiments.

9.1 Fonts

The fonts can usually be changed wholesale via the font key at the axis level, and for individual elements by specifying the font for that element using the keys in Reference 9.1.

Reference 9.1: Fonts in plots

\begin{axis}[*AxisOptions*] *PlotCommands* \end{axis}

Axis option	Description
font=*Font*	Default font
title style={*Style*}	Append to every title style; see also §10.9
legend style={*Style*}	Append to every axis legend; see also §9.4.3
xy label style={*Style*}	Append to every axis xy label; see also §10.10
xy ticklabel style={*Style*}	Append to every xy tick label

Style	Description
every title style	Title font, location, etc.; see §9.1, §10.9
every axis legend	Legend font, location, etc.; see §9.1, §9.4.3
every axis xy label	Label font, locations, etc.; see §9.1, §10.10
every xy tick label	Tick label font, number formats, etc.; see §9.1

The font=*Font* command at the axis level sets all the fonts as specified. So, for instance, setting font=\small\sffamily universally changes the fonts to the specified size and style.

We can change specific fonts inside the appropriate style, overriding the axis-level specification. For instance, `title style={font=\bfseries\sffamily\large}` sets the title font. Note that the `font` key is completely reset, as the effect of multiple font keys is not cumulative.

Example 9.1 changes the default font to `font=\footnotesize`. It also specifies `title style={font=\bfseries}`, which entirely resets the font for the title, with a side effect of switching the size back to the default of `\normalsize`. It uses `data.dat` (on page 55).

Example 9.1: Changing fonts

```
\begin{tikzpicture}
  \begin{axis}[scale only axis,
      width=3.5cm,height=3cm,
      font=\footnotesize,
      title style={font=\bfseries},
      title=Line Plot, legend entries={A},
      xlabel=Input, ylabel=Output,
    ]
    \addplot[blue,mark=*] table{data.dat};
  \end{axis}
\end{tikzpicture}
```

The default fonts for every axis can be set by adding something like what is shown in Reference 9.2 to the preamble.

Reference 9.2: Example default size and font setup

```
\pgfplotsset{
  every axis/.append style={
    title style={font=\bfseries},
    label style={font=\small},
    ticklabel style={font=\footnotesize},
    legend style={font=\small}
}}
```

The title, axis labels, and tick labels are text nodes. The legend node is a little more complicated, but it is also a node. Therefore, we can use certain node options from Chapter 3 to modify their appearance. We defer most details until the later discussion of nuances (see Chapter 10), but mention here that useful options include `rotate`, `xshift`, `yshift`, `anchor`, `inner sep`, and more. Additionally, using `draw=red` to outline the node can also be useful in debugging the alignment of the nodes.

9.2   Number formatting and scaling of tick labels

If the tick labels are not appearing exactly as you like, they can be manually specified (as strings) using `xticklabels` and `yticklabels`. Alternative options are provided in Reference 9.3. For the numeric tick labels, we can specify the format using PGF number-formatting capabilities. Additionally, PGFPLOTS automatically scales larger numbers. This can be modified using the `scaled ticks` option.

> ### Reference 9.3: Tick number formatting
>
> `\begin{axis}[`*AxisOptions*`]` *PlotCommands* `\end{axis}`
>
> | Axis option | Description |
> |---|---|
> | [xy]`ticklabel style=`*Style* | Append to `every` [xy] `tick label` |
> | `log ticks with fixed point` | Write out numbers on log axis |
> | `scaled` [xy] `ticks=`*ScalingOption* | Control tick scaling |
>
> | Scaling option | Description |
> |---|---|
> | `true` | Factor out common exponents for linear axes |
> | `false` | No scaling |
> | `base 10:`*E* | Scale axes by 10^{-E} |
> | `real:`*R* | Scale axes by R |
>
> | Style | Description |
> |---|---|
> | `every` [xy] `tick label` | Style for tick labels |
> | `log plot exponent style` | Style for exponents in log ticks |
>
> | Number format | Description |
> |---|---|
> | `/pgf/number format/fixed` | Fixed precision (e.g., not scientific format) |
> | `/pgf/number format/precision=`*N* | Maximum number of digits following decimal |
> | `/pgf/number format/zerofill` | Fill decimal with zeros to specified precision |
> | `/pgf/number format/1000 sep={}` | No commas in numbers greater than 1,000 |

The number formatting for *linear* axes can be specified via [xy]`ticklabel style`, which appends to the style `every` [xy] `tick label`. The number formatting for *log* axes is not specific to the x- or y-axis and is controlled via the style `log plot exponent style`. The format of the numbers can be specified via various `/pgf/number format/` keys, and we list some of the most useful number formats above. A full set of options can be found under "Number Printing" in the PGF/TikZ manual.

Since typing out `/pgf/number format` repeatedly can be tiresome, we suggest the shortcuts in Reference 9.4. The basic idea is that `numformat={fixed,precision=2}` expands to `/pgf/number format/fixed`, `/pgf/number format/precision=2`.

> ### Reference 9.4: Number formatting shortcut
>
> ```
> \pgfplotsset{
>   numformat single/.style={/pgf/number format/#1},
>   numformat/.style={numformat single/.list={#1}}
> }
> ```

Numeric tick values are scaled automatically by PGFPLOTS. This can be disabled via `scaled ticks=false`. It is also possible to specify the scaling manually. The placement of the scaling can be awkward, as we will see in the example that follows. I generally recommend to disable automatic scaling and instead handle it manually using `y expr` and then specify the units in the axis label, as we did, for instance, in Example 8.1.

In Example 9.2 using `res.dat` (on page 73), we give examples of tick number formatting. The x-tick labels would be printed as "2,018", "2,019", and so on if we did not specify `1000 sep={}`. The y-tick labels are scaled automatically by 10^4, and the scaling indicator is above the upper left corner of the axes. The y-tick labels would be "3" rather than "3.0" and "4" rather than "4.0" if we did not specify `{precision=1,zerofill}`. The option `title style` resets the title font (overriding `font=\footnotesize`) and shifts the title up by 2 mm to avoid overwriting the scaling.

Example 9.2: Tick number formatting

```
\pgfplotsset{
  numformat single/.style={/pgf/number format/#1},
  numformat/.style={numformat single/.list={#1}}}
\begin{tikzpicture}
  \begin{axis}[scale only axis, width=6cm, height=4cm,
      font=\footnotesize, title=Del Valle Reservoir,
      xticklabel style={numformat={1000 sep={}}},
      yticklabel style={numformat={precision=1, zerofill}},
      title style={font=\bfseries,yshift=2mm} ]
    \addplot[blue,mark=*] table[y=max af]{res.dat};
    \addplot[orange,mark=*] table[y=min af]{res.dat};
  \end{axis}
\end{tikzpicture}
```

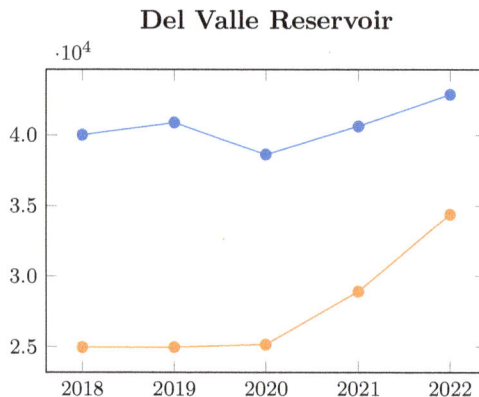

9.3 Tick marks and grids

In addition to specifying the location of the ticks and their labels, we can also control their appearance using the options described in Reference 9.5.

Display of major and/or minor ticks is controlled by `ticks` or axis-specific options like `xminorticks`. Minor ticks may appear automatically on a logarithmic axis such as in Example 7.7 on page 62. Minor ticks can be added with the option `minor [xy]tick num`, as we will demonstrate later in Example 9.4. The length of the major and minor ticks is controlled by `major tick length` and `minor tick length`, respectively. The style of the ticks (which are drawn as lines) is appended to by `tick style`. The orientation

> **Reference 9.5: Tick marks and grids**
>
> \begin{axis}[*AxisOptions*] *PlotCommands* \end{axis}
>
> | Axis option | Description |
> |---|---|
> | ticks=*Option* | Show ticks: minor, major, both, or none |
> | xymajorticks=*TF* | Show major ticks on axis: true or false |
> | xyminorticks=*TF* | Show minor ticks on axis: true or false |
> | minor xy tick num=*Value* | Number of minor ticks between major ticks |
> | major tick length=*Value* | Length of major tick, defaults to 0.15 cm |
> | minor tick length=*Value* | Length of minor tick, defaults to 0.10 cm |
> | tick style=*Style* | Append to every tick |
> | tick align=*Align* | Options are inside, outside, center |
> | grid=*Option* | Show grids: minor, major, both, or none (default) |
> | xymajorgrids=*TF* | Show major grid lines on axis, default false |
> | xyminorgrids=*TF* | Show minor grid lines on axis, default false |
> | grid style=*Style* | Append to every axis grid |
>
> | Axis style | Description |
> |---|---|
> | every tick | Formatting of ticks; initially {very thin, gray} |
> | every axis grid | Formatting of grid; initially {thin, black!25} |

of the ticks is set by tick align; this defaults to inside but changes automatically to outside for bar plots.

Display of major or minor grids is controlled by grid (not plural), or axis-specific options like ymajorgrids (plural).

> **?** *Oddity:* It is not clear why grid is singular when all the analogous options are plural.

In Example 9.3, using **runtime.csv** (on page 68), horizontal grids are enabled via ymajorgrids=true, minor y tick marks are disabled via yminorticks=false, and the major y ticks are still there but are not seen because their length is set to zero via major tick length=0em.

> **Example 9.3: Showing a grid**
>
> ```
> \pgfplotsset{table/col sep=comma}
> \begin{tikzpicture}
>   \begin{semilogyaxis}[scale only axis,
>       width=3in, height=1in, major tick length=0em,
>       every axis plot post/.style={fill,draw=black},
>       ybar, bar width=8pt,
>       ymajorgrids=true, yminorticks=false, ylabel=Run time (sec),
>       xtick=data, enlarge x limits=0.2, xlabel=Dataset,
>       legend entries={Method 1, Method 2, Method 3, Method 4},
> ```

```
    legend pos=outer north east,
    legend style={font=\footnotesize}
  ]
  \addplot[blue] table[y=M1] {runtime.csv};
  \addplot[orange] table[y=M2] {runtime.csv};
  \addplot[gray] table[y=M3] {runtime.csv};
  \addplot[yellow] table[y=M4] {runtime.csv};
  \end{semilogyaxis}
\end{tikzpicture}
```

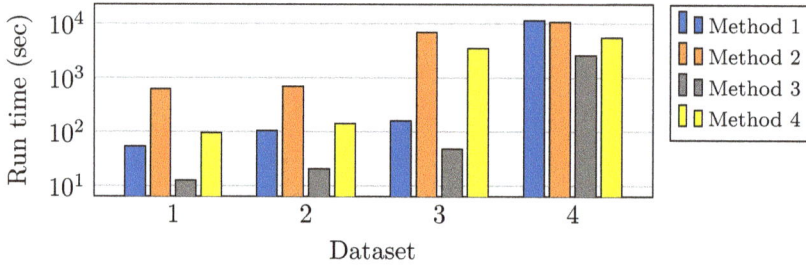

9.4   Legend refinements

The keys in Reference 9.6 are useful for refining the style of the legend.

Reference 9.6: Legend refinements

\begin{axis}[*AxisOptions*] *PlotCommands* \end{axis}

Axis option	Description
legend entries={*E1*,...,*EN*}	Specify legend entries
legend pos=*Pos*	Options are south west, south east, north west, north east (default), and outer north east
legend columns=*N*	Default is 1; use -1 for horizontal legend
legend cell align=*Align*	Options: left, right, or center (default)
legend style={*Style*}	See below for style options
reverse legend	Reverses the order of the legend entries

Plot option	Description
forget plot	Exclude plot from legend

Legend style option	Description
at={(*X*,*Y*)}	Location of legend node (alternative to legend pos)
anchor=*Anchor*	Anchor of legend node
draw=*Color*	Set legend outline color or none for no outline
fill=*Color*	Set legend fill color or none for no fill
font=*Font*	Font style for legend text

The legend entries are specified as the axis key legend entries as in Example 9.3. If there are n legend entries and $p < n$ plots, then only the first p legend entries are used. Conversely, if $p > n$, only the first n plots are included in the legend. The key forget plot in an \addplot command entirely ignores the plot with respect to the legend. The legend entries key should be sufficient for most purposes, but there are other ways to declare legends, such as \legend and \addlegendentry; see the PGFPLOTS manual for details.

Under the hood, the legend uses the TikZ matrix environment, which we do not cover but mention mainly so that the key legend cell align has some context. It refers to the alignment of cells within the matrix. Additionally, the text nodes within the matrix are changed via legend style={nodes={Style}}. We refer the user to the PGFPLOTS manual for further details.

9.4.1 Legend position

The position of the legend can be set in one of two ways. The legend pos axis key can use one of five predefined settings: south west, south east, north west, north east (default), and outer north east. Alternatively, the position can be set manually using options at and anchor *inside* the legend style key. The default legend position of north east is equivalent to legend style={at={(0.98,0.98)}, anchor=north east}.

In Example 9.4, we use legend style={at={(1.03,0.5)},anchor=west} to place the legend to the outside right middle of the axes. The point $(1.03, 0.5)$ is with respect to the axis description cs, which is the coordinate system that places $(0,0)$ at the lower left of the axes and $(1,1)$ at the upper right (see Section 10.4 for further details). This example uses data.dat (on page 55).

Example 9.4: Positioning legend using at

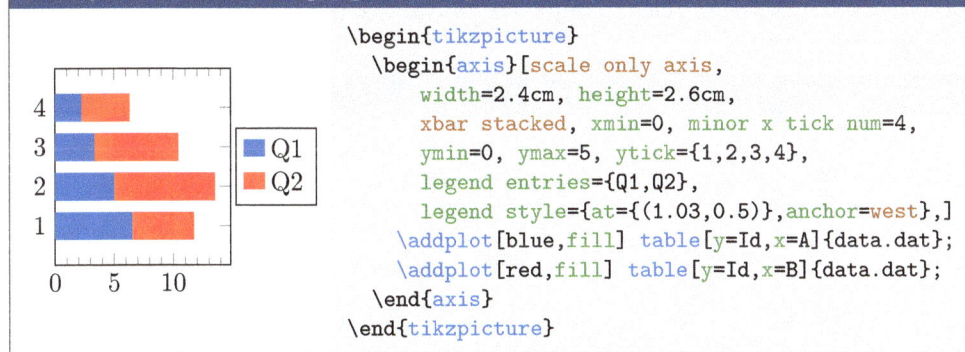

```
\begin{tikzpicture}
  \begin{axis}[scale only axis,
    width=2.4cm, height=2.6cm,
    xbar stacked, xmin=0, minor x tick num=4,
    ymin=0, ymax=5, ytick={1,2,3,4},
    legend entries={Q1,Q2},
    legend style={at={(1.03,0.5)},anchor=west},]
    \addplot[blue,fill] table[y=Id,x=A]{data.dat};
    \addplot[red,fill] table[y=Id,x=B]{data.dat};
  \end{axis}
\end{tikzpicture}
```

9.4.2 Horizontal legend

By default, the legend entries are placed in a single column, but we can instead specify the number of columns using legend columns. A horizontal legend can be achieved by setting legend columns=-1. The default PGFPLOTS formatting generally leaves insufficient space between the legend entries in the horizontal format, so manual spacing (e.g., via ~) is recommended in the declaration.

In Example 9.5, we show a horizontal legend, achieved via legend columns=-1. We manually add a bit of space after the first legend entry using explicit spaces (~~~) in the `legend entries` setting. The placement is set in the `legend style` to have its south anchor point at coordinate $(0.5, 1.03)$, meaning that it's at the halfway point on the x-axis and slightly above the top of the y-axis. We also set the legend font to `\large` using `legend style`. This example uses `data.dat` (on page 55).

Example 9.5: Horizontal legend

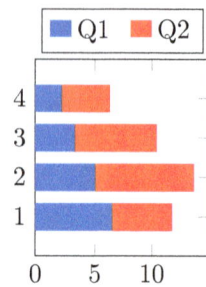

```
\begin{tikzpicture}
  \begin{axis}[scale only axis,
      width=2.4cm, height=2.6cm,
      xbar stacked, xmin=0,
      ymin=0, ymax=5, ytick={1,2,3,4},
      legend columns=-1, legend entries={Q1~~,Q2},
      legend style={at={(0.5,1.03)},anchor=south} ]
    \addplot[blue,fill] table[y=Id,x=A]{data.dat};
    \addplot[red,fill]  table[y=Id,x=B]{data.dat};
  \end{axis}
\end{tikzpicture}
```

9.4.3   Legend style

Other options can also be specified via `legend style`, using keys such as `draw`, `fill`, and `font`. For instance, the box around the legend is omitted by setting `legend style={draw=none}`. (We discussed the `font` option in Section 9.1.)

In Example 9.6, we show how to create a semi-transparent legend. The legend appears in the north east corner, which is the default `legend pos`. We specify the legend entries using `legend entries`. We modify the legend style using `legend style` and set the `fill opacity` on the legend to 0.7, rendering it semi-transparent. By default, the `text opacity` inherits the `fill opacity`, so we explicitly set `text opacity=1` to override that inheritance. This example uses `data.dat` (on page 55).

Example 9.6: Semi-transparent legend

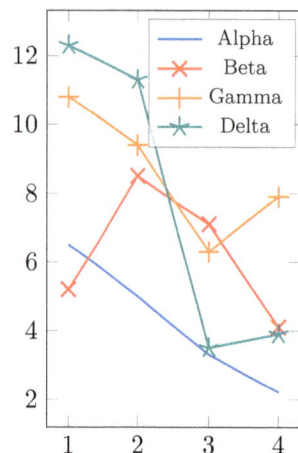

```
\pgfplotstableread{data.dat}{\data}
\begin{tikzpicture}
  \begin{axis}[
      scale only axis, width=3.5cm,height=5.5cm,
      every axis plot/.style={thick,mark
    size=4pt},
      legend entries={Alpha,Beta,Gamma,Delta},
      legend style={ultra thin,fill opacity=0.7,
        text opacity=1,font=\footnotesize}]
    \addplot[blue] table[y=A]{\data};
    \addplot[red,mark=x] table[y=B]{\data};
    \addplot[orange,mark=+] table[y=C]{\data};
    \addplot[teal,mark=star] table[y=D]{\data};
  \end{axis}
\end{tikzpicture}
```

9.5   Aligning multiple axes using group plots

If you want to have a group of aligned plots, the simplest way to do so is using the groupplots library as documented in Reference 9.7. A groupplot environment replaces the usual axis environment, and a new axis within the group plot is demarcated by \nextgroupplot, which takes additional optional axis options. Each axis in the group plot can have one or more \addplot commands.

The arguments to the groupplot environment specify the axis options that are shared by all the plots in the group. Most importantly, the special group style contains the group options. The options can alternatively be listed individually preceded by group/. The layout of the group plot is specified within group style using group size=*M* by *N*. There are options for the vertical and horizontal spacing and for sharing axis labels and tick labels.

Reference 9.7: Aligning plots with groupplot

```
Preamble:\usepgfplotslibrary{groupplots}
\begin{tikzpicture}[...]
  \begin{groupplot}[
      group style={group size=Cols by Rows,GroupOptions},
      AxisOptions
  ]
    \nextgroupplot[AxisOptions]
    \addplot ...;
    \nextgroupplot[AxisOptions]
    \addplot ...;
    ...
  \end{groupplot}
\end{tikzpicture}
```

Group option	Description
horizontal sep=*H*	Horizontal separation between plots, default 1 cm
vertical sep=*V*	Vertical separation between plots, default 1 cm
xlabels at=*Loc*	Label location, either all (default) or edge bottom
xticklabels at=*Loc*	Same options as xlabels at for the x-axis tick labels
ylabels at=*Loc*	Label location, either all (default) or edge left
yticklabels at=*Loc*	Same options as ylabels at for the y-axis tick labels
group name=*Name*	Name of group plot, allowing reference to subplots using *Name*.c*C*r*R* where *C* is column index and *R* is row index

---

?₂? Oddity: The environment is groupplot (singular), and the library is groupplots (plural).

In Example 9.7, we provide an example of using the groupplot environment using data.dat (on page 55). We have four plots arranged in a 2×2 grid. We specify both the horizontal and vertical distance between plots. The plots share their size, axis limits, and axis labels. We avoid repeating the labels for every graph using the key xlabels at=edge bottom and ylabels at=edge left. Each axis has its own title, specified as an option to \nextgroupplot.

Example 9.7: Aligning plots with `groupplot`

```
Preamble:\usepgfplotslibrary{groupplots}
\begin{tikzpicture}
  \begin{groupplot}[
      group style={
        group size=2 by 2, horizontal sep=1cm, vertical sep=1.5cm,
        xlabels at=edge bottom, ylabels at=edge left,
      },
      scale only axis, width=4cm, height=2cm,
      xlabel=$x$, xmin=0.5,xmax=4.5,xtick distance=1,
      ylabel=$y$, ylabel near ticks,
      ymin=0, ymax=15, ytick distance=3
    ]
    \nextgroupplot[title=Dataset A]
    \addplot[blue,mark=*] table[y=A]{data.dat};
    \nextgroupplot[title=Dataset B]
    \addplot[red,mark=*] table[y=B]{data.dat};
    \nextgroupplot[title=Dataset C]
    \addplot[orange,mark=*] table[y=C]{data.dat};
    \nextgroupplot[title=Dataset D]
    \addplot[teal,mark=*] table[y=D]{data.dat};
  \end{groupplot}
\end{tikzpicture}
```

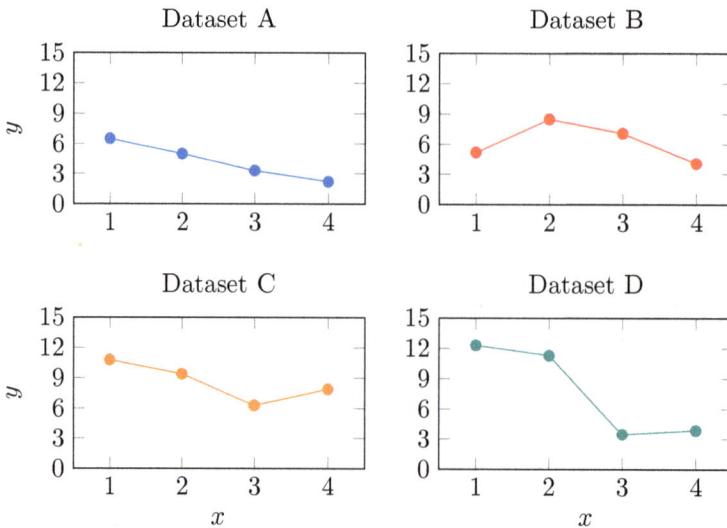

9.6 Fill between two plots

Reference 9.8 explains how to fill the space between two plots. This requires an extra library in the preamble: \usepgfplotslibrary{fillbetween}. The first step is to name the plots using the name path=*Name* option. The second step is to use the special fill between plot. Its required input argument is of=*A* and *B*. It has also an optional second input argument to specify the domain, i.e., soft clip={domain=*X1*:*X2*}, meaning that the fill is limited to x-values between $x = X1$ and $x = X2$.

Reference 9.8: Filling between two plots

```
\addplot[name path=A] ...;
\addplot[name path=B] ...;
\addplot[fill=Color] fill between [of=A and B, FillBetweenOpts];
```

Fill between option	Description
soft clip={domain=*Xmin*:*Xmax*}	Domain for fill between is [*Xmin*,*Xmax*]

Example 9.8 demonstrates fill between using **res.dat** (on page 73). The first plot is red and named via name path=**max**. The second plot is blue and named via name path=**min**. The third plot fills between the first two, referencing them via fill between [of=**min** and **max**]. The legend shows both plots and the fill between.

Example 9.8: Fill between

```
Preamble:\usepgfplotslibrary{fillbetween}
\begin{tikzpicture}
  \begin{axis}[scale only axis, width=6cm, height=3cm,
      xticklabel style={/pgf/number format/1000 sep=},font=\small,
      ymin=0, legend pos=south east,
      legend entries={max,min,difference}
  ]
    \addplot[red,name path=max]  table[y=max af]{res.dat};
    \addplot[blue,name path=min] table[y=min af]{res.dat};
    \addplot[fill=yellow] fill between [of=max and min];
  \end{axis}
\end{tikzpicture}
```

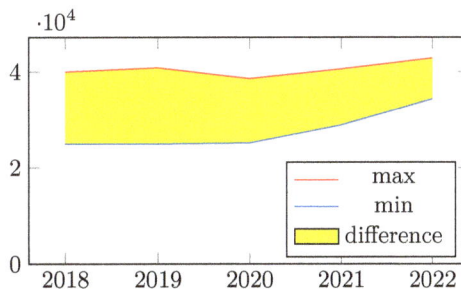

In Example 9.9, we demonstrate the soft clip={domain=*X1*:*X2*} option. The first plot is from **data.dat** (on page 55), except that we add name path=**mydata**. The second plot is an artificial plot of sorts, just adding an invisible line (draw=none) at zero with an inline table. This plot is named via `name path`=`zero`. Finally, the `fill between` adds the option `soft clip`=`{domain=1.5:3}` to limit the fill-between operation to the x-values ranging from 1.5 to 3.

Example 9.9: Fill between with clipped domain

```
Preamble:\usepgfplotslibrary{fillbetween}
\begin{tikzpicture}
  \begin{axis}[scale only axis, width=6cm, height=3cm]
    \addplot[name path=mydata,blue,mark=*] table{data.dat};
    \addplot[name path=zero,draw=none]
      table[col sep=comma,row sep=\\]{1,0\\4,0\\};
    \addplot [fill=gray!50] fill between[of=mydata and zero,
      soft clip={domain=1.5:3}];
  \end{axis}
\end{tikzpicture}
```

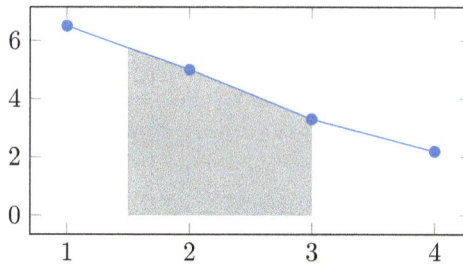

Example 9.10 shows a more complicated example. We have a data file, `hoeffding.csv` (on the current page), for which we only show a few lines. The column `num samples` is the cumulative number of samples from a uniform distribution on [2,4], the column `sample mean` is the statistical sample mean (usually denoted as $\bar{\mu}$), and the column `confidence interval` is the statistical confidence interval about the true mean ($\mu = 3$), which contains the sample mean with confidence at least 90% according to Hoeffding's inequality in statistics.

Data File: hoeffding.csv

```
num samples,sample mean,confidence interval
1,3.6294,2.4477
2,3.7205,1.7308
3,3.2317,1.4132
```
many more lines omitted...

In all the `\addplot` commands, the x-values are not specified and default to the first column (`num samples`). The first `\addplot` plots `sample mean`. The second `\addplot` plots a horizontal line at $y = 3$ using an inline table that has just two points: $(1, 3)$ and $(100, 3)$. The third `\addplot` plots the upper confidence interval, using `y expr`=`3+\thisrow{confidence interval}`. It also uses the style `cib`, defined in the axis options, and sets the name of the path via `name path`=`upper`. The `cib` style includes `forget plot` so this plot is not included in the legend. The fourth `\addplot` adds the lower confidence internal, and names it via `name path`=`lower`. The fifth and final `\addplot` fills between the upper and lower plots. Observe that the fill between occurs in the background and does not overwrite the other plots. It can, however, overlap the axes, so we include axis on top in the axis options.

Example 9.10: Fill between

```
Preamble:\usepgfplotslibrary{fillbetween}
\pgfplotsset{table/col sep=comma}
\begin{tikzpicture}
  \begin{axis}[
     scale only axis,
     width=8cm, height=3cm, ymin=2, ymax=4, enlarge x limits=false,
     xlabel={Cumulative Number of Samples},
     legend pos=outer north east, axis on top,
     legend entries={{Sample mean, $\bar \mu$},{Mean, $\mu$},
       90\% Confidence},
     cib/.style={green,dashed,thick,forget plot} % Custom style
  ]
  \addplot[blue,mark=*,mark size=1pt]
    table[y=sample mean] {hoeffding.csv};
  \addplot[red, thick] table[row sep=\\]{1,3\\100,3\\};
  \addplot[cib, name path=upper]
    table[y expr=3+\thisrow{confidence interval}] {hoeffding.csv};
  \addplot[cib, name path=lower]
    table[y expr=3-\thisrow{confidence interval}] {hoeffding.csv};
  \addplot[fill=green!5] fill between [of=upper and lower];
  \end{axis}
\end{tikzpicture}
```

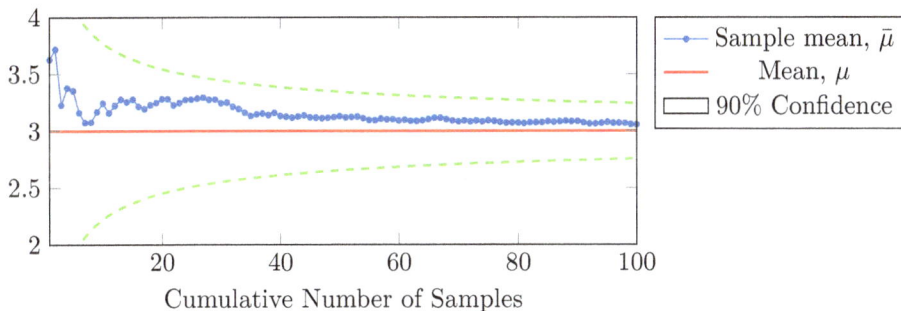

Chapter 10

Advanced Topics and Nuances

10.1 Symbolic coordinates

There are a couple of different ways to plot data sets that contain non-numeric (i.e., symbolic) data values. The easiest method is to map the symbolic data to numeric values via external preprocessing and then use xticklabels to display the original symbolic values on the axis. In Example 10.1 using `runtime.csv` (on page 68), we use x=Num to specify that the x-values are given by the numeric values in the column named Num, but we overwrite the labels via xticklabels={A,B,C,D} in the axis options.

Example 10.1: Changing tick labels to be symbolic

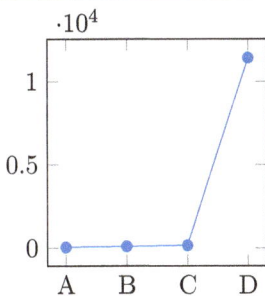

```
\begin{tikzpicture}
  \begin{axis}[table/col sep=comma,
      scale only axis, width=3cm, height=3cm,
      xtick={1,2,3,4}, xticklabels={A,B,C,D}]
    \addplot[blue,mark=*]
      table[x=Num,y=M1]{runtime.csv};
  \end{axis}
\end{tikzpicture}
```

Alternatively, PGFPLOTS provides commands to create plots directly from non-numeric data without having to map the symbols to their underlying numeric values in a preprocessing step.

Reference 10.1: Setting symbolic coordinates

\begin{axis}[*AxisOptions*] ... \end{axis}

Axis option	Description
symbolic xy coords={S1,S2,...,SN}	Set symbolic coordinates and order
xtick=data	Recommended with above

91

Using symbolic coordinates requires symbolic x coords={S1,S2,...,SN}, which provides the ordering of the symbols. Under the hood, it maps the symbols {S1,S2,...,SN} to the numeric values $\{0,1,...,N-1\}$. In Example 10.2, we use the x=Ltr to specify that the x-values are given by the letters in the column with the heading Ltr. We then specify the ordinal order via symbolic x coords={A,B,C,D}. We also specify xtick=data to ensure that the tick marks are correct since symbolic coordinates can sometimes produce odd ticks in which the symbolic tick labels are repeated.

Example 10.2: Direct symbolic coordinates

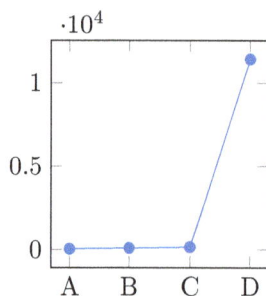

```
\begin{tikzpicture}
  \begin{axis}[table/col sep=comma,
    scale only axis, width=3cm, height=3cm,
    symbolic x coords={A,B,C,D},
    xtick=data]
    \addplot[blue,mark=*]
      table[x=Ltr,y=M1]{runtime.csv};
  \end{axis}
\end{tikzpicture}
```

10.2   Axis refinements: direction, mode, equal units

There are several options shown in Reference 10.2 that can be useful in modifying the axis direction, mode, and sizing.

Reference 10.2: Setting axis size and limits

\begin{axis}[*AxisOptions*] ... \end{axis}

Axis option	Description
xy dir=reverse	Reverse the axis direction
xy mode=*Option*	Set mode of axis to linear or log
axis equal	Equalize axis units, potentially changing limits
axis equal image	Equalize axis units, potentially changing size

---

10.2.1   Reversing the axis direction

The axes go from smallest to largest by default, but we can reverse the direction using the x dir=reverse and y dir=reverse keys. This is a nice alternative to manually adjusting the plot coordinates and tick labels to achieve such an effect. In Example 10.3, we reverse the y-axis direction so that the numbers read $1,2,3,4$ from top to bottom rather than the default of $4,3,2,1$.

> **Example 10.3: Reversing y-axis direction**
>
> ```
> \begin{tikzpicture}
>   \begin{axis}[xbar stacked,
>     scale only axis,width=3cm, height=3cm,
>     ymin=0,ymax=5,ytick={1,2,3,4},y dir=reverse,
>     legend entries={A,B}, legend pos=south east]
>       \addplot[blue,fill] table[y=Id,x=A]{data.dat};
>       \addplot[red,fill]  table[y=Id,x=B]{data.dat};
>   \end{axis}
> \end{tikzpicture}
> ```

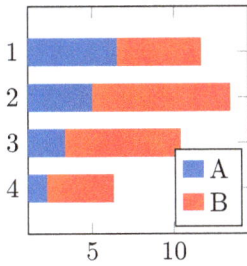

10.2.2 Changing the axis mode to log or linear

Recall that the axis command has four variants (axis, semilogxaxis, semilogyaxis, and loglogaxis) to specify which axes will be logarithmic. We can alternatively specify the modes of the axes as options; e.g.,

```
\begin{axis}[xmode=log] ... \end{axis}
```

is equivalent to

```
\begin{semilogxaxis} ... \end{semilogxaxis}
```

In Example 10.4, we use a regular axis environment and then set the modes to be logarithmic.

> **Example 10.4: Specifying log plot via axis option**
>
> ```
> \pgfplotsset{table/col sep=comma}
> \begin{tikzpicture}
>   \begin{axis}[scale only axis, width=4cm,
>     height=3cm, xmode=log, ymode=log]
>       \addplot[mark=o]
>       table[y=sp_50]{compress.csv};
>   \end{axis}
> \end{tikzpicture}
> ```

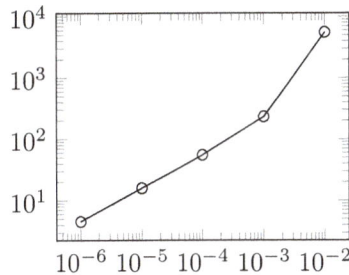

10.2.3 Equalizing axes

We can equalize the units on the axes automatically using axis equal or axis equal image keys, which have slightly different effects. In Example 10.5, we show the effect of these keys. In the default plot, the unit sizes on the x- and y-axes are different. The range of indices on the y-axis is larger, so the x-axis is modified by the options. Using the key axis equal enlarges the x-axis limits so that the unit sizes are equal to those of the y-axis; it does not change the width or height. Conversely, the key axis equal image does not change the axis limits but instead changes the width. In both cases, one unit on the x-axis is the same size as one unit on the y-axis.

Example 10.5: Equalizing axes

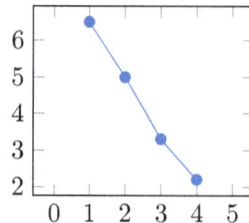

```
\pgfplotsset{
  every axis/.append style={scale only axis,width=3cm, height=2.5cm,
    xtick distance=1, ytick distance=1}
}
\begin{tikzpicture}
  \begin{axis}[title=default (fit axes to data)]
    \addplot[blue,mark=*] table{data.dat};
  \end{axis}
\end{tikzpicture}
~~~~ % spaces
\begin{tikzpicture}
  \begin{axis}[axis equal, title={\texttt{axis equal}}]
    \addplot[blue,mark=*] table{data.dat};
  \end{axis}
\end{tikzpicture}
~~~~ % spaces
\begin{tikzpicture}
  \begin{axis}[axis equal image, title={\texttt{axis equal image}}]
    \addplot[blue,mark=*] table{data.dat};
  \end{axis}
\end{tikzpicture}
```

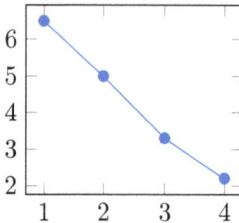

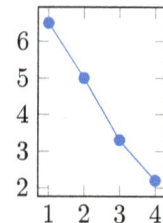

10.3   Plotting mathematical expressions

It is possible to plot math functions using PGFPLOTS, i.e., $y = f(x)$ where $f(x)$ is an expression in a single variable, x. This should be used with some caution, as there are many scenarios where it is more efficient and accurate to calculate the values externally and save them to a table. Nevertheless, the ability to quickly plot a function can be useful, so we cover it here in Reference 10.3. The general format for a math expression substitutes expression where we would normally have table in the \addplot command. The math expressions are the same as those used in other math operations within TikZ and as described in Section 4.4.

Reference 10.3: addplot

\addplot[*PlotOptions*] expression[*MathOptions*] {*MathExpression*};

Math option	Description
domain=*Xmin*:*Xmax*	Sample domain [*Xmin*,*Xmax*]; default is $[-5, 5]$
samples=*N*	Use *N* evenly spaced samples; defaults to 25
samples at={*X1,X2,...,XN*}	Specific points; overrides domain and samples
variable=*Letter*	Specifies variable name; defaults to x

There are several arguments for an expression. We can specify the domain and number of samples using domain and samples. By default, samples are evenly spaced. For a log x-axis, they are evenly spaced with respect to the log scale. Alternatively, we can specify the precise sample points using samples at. PGFPLOTS assumes the expression is a function in terms of x, but we can change the variable via variable. For example, we could write a function in terms of t as follows.

\addplot[...] expression[variable=t] {t^2-2*t+4};

In Example 10.6, we give a simple example of using a mathematical expression. The \addplot command is followed by expression rather than table. The domain is specified as $[-4, 4]$ using domain=-4:4. The number of samples is 9 via samples=9. For instance, we would have gotten the exact same samples using samples at={-4,-3,...,4}. The mathematical expression is $\sqrt{e^x} - x$, using sqrt(exp(x))-x.

Example 10.6: Plotting a math expression

$$y = \sqrt{e^x} - x$$

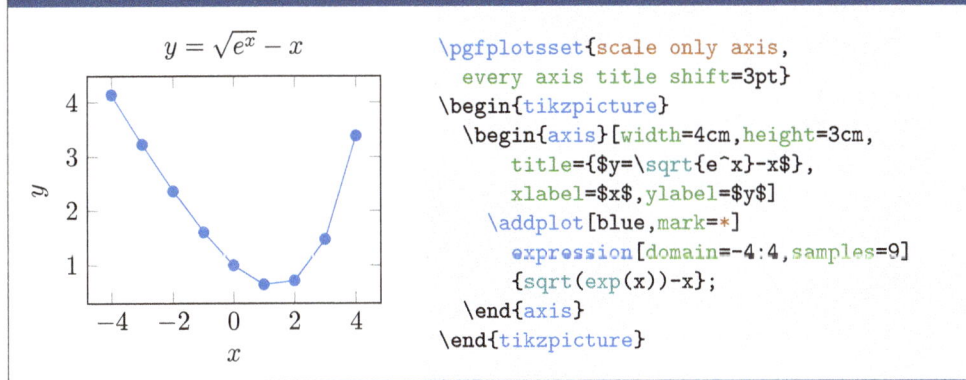

```
\pgfplotsset{scale only axis,
    every axis title shift=3pt}
\begin{tikzpicture}
    \begin{axis}[width=4cm,height=3cm,
        title={$y=\sqrt{e^x}-x$},
        xlabel=$x$,ylabel=$y$]
        \addplot[blue,mark=*]
            expression[domain=-4:4,samples=9]
            {sqrt(exp(x))-x};
    \end{axis}
\end{tikzpicture}
```

10.4 TikZ commands within an axis environment

TikZ commands can be used inside an axis environment to draw additional lines and nodes. However, care is needed in drawing with respect to the appropriate coordinate system; see Reference 10.4. We have to specify axis cs to place a point relative to the plot axes, and relative coordinates have their own special coordinate system, axis direction cs. Additionally, any annotations that extend outside of the axes are automatically clipped, though this can be disabled with clip=false. Nevertheless, to avoid problems with clipping, I generally recommend creating named coordinates inside the axis to mark a location and then subsequently use these coordinates for drawing paths and nodes after the plot is drawn.

Reference 10.4: PGFPLOTS coordinate systems

```
\path[...] (axis cs:X,Y) -- ++(axis direction cs:DX,DY) ...;
\path[...] (rel axis cs:X,Y) ...;
```

Coordinate system	Description
axis cs	Axis coordinates, but not for relative coordinates!
axis direction cs	For relative coordinates, combined with axis cs
rel axis cs	Maps $(0,0)$ to lower left and $(1,1)$ to upper right
axis description cs	For 2D graphs, the same as rel axis cs

⚠ **Remember:** Use axis cs to place a point with respect to the numbers on the axes, but use axis direction cs to place a point relative to another point.

Example 10.7: Combining TikZ and PGFPLOTS

```
\begin{tikzpicture}[
    arw/.style={<-,shorten <=2mm,shorten >=4mm,thick},
    lbl/.style={draw,fill=white,fill opacity=0.8}
  ]
  \begin{axis}[scale only axis,width=6cm,height=2cm,xmin=1.5]
    \addplot[mark=*] table{data.dat};
    \draw[red,arw] (axis cs:2,5) -- ++(axis direction cs:0,-1.5)
       node[lbl] {$(2,5)$};
    \draw[blue,arw] (axis cs:3,3.3) -- (rel axis cs:0.5,1)
       node[lbl] (A) {$(3,3.3)$};
    \coordinate (B) at (axis cs:4,2.2);
  \end{axis}
  \draw[orange,arw] (B) -- (B|-A) node[lbl] {$(4,2.2)$};
\end{tikzpicture}
```

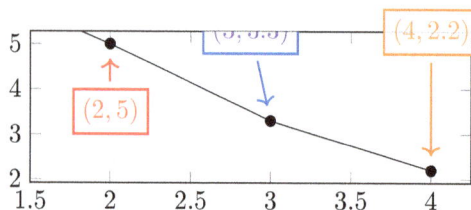

In Example 10.7, we use PGFPLOTS commands to plot `data.dat` (on page 55). In this case, we set xmin=1.5, so the plot goes outside the axes and is clipped. We then augment the plot by using TikZ commands. The first \draw command adds a red arrow starting at (axis cs:2,5), which means it starts at the coordinate $(2,5)$ relative to the coordinate axes. The next point is relative to the first within the axes, using ++(axis direction cs:0,-1.5). The second \draw command adds a blue arrow starting at (axis cs:3,3.3), so this means at the point $(3,3.3)$ relative to the axes. The arrow goes to (rel axis cs:0.5,1), which is the point halfway along the top axis. It puts a node there, but observe that the node is cut off where it crosses the axis. We could fix this with clip=false as an axis option, but this would have the side effect of not clipping the plot. Instead, the last Ti*k*Z command within the axis environment places a coordinate at (axis cs:4,2.2), which is the point (4, 2.2) relative to the axis coordinate system. Then, the arrow and node are finally drawn *outside* the axis environment, and so the node is not clipped with the axis.

As a final note, we mention that PGFPLOTS has an option nodes near coords that labels data points with their *y*-values. It can also be used in other ways, and we refer to the PGFPLOTS manual for details.

10.5   Axis placement and reference points

An axis environment is a special Ti*k*Z node, and there are special anchors that can be used for placing it inside the tikzpicture environment. This can be especially helpful when aligning multiple axes. Reference 10.5 recalls relevant node options for placing an axis, and Fig. 10.1 on the following page details the predefined anchor points. The compass point anchors (north east, etc.) shown in orange are on the axes themselves. The outer compass point anchors (outer east, etc.) shown in green expand to contain the tick marks, labels, and legend. Finally, the reference compass point anchors (left of south west, etc.) shown in blue project the inner compass points to the outer border. The anchor=*Anchor* command specifies where to anchor an axis, and the default is south west. The option at=(*C*) will place the anchor at coordinate *C*.

Reference 10.5: Axis name and anchor

\begin{axis}[*AxisOptions*] *PlotCommands* \end{axis}

Axis option	Description
name=*Name*	Name the axis node
at=(*C*)	Coordinate for placement of the axes; default is (0, 0)
anchor=*Anchor*	Anchor the axis node; default south west

In Example 10.8, we show an example of placing an axis at a specific coordinate and with a specific anchor. For reference, we draw a Ti*k*Z grid and place coordinate (A) at (3,3). Then we create an axis environment with the axis options at=(A),anchor=center. This means that the center anchor of the axes is placed at coordinate (A).

Example 10.8: Placing an axis with respect to a coordinate

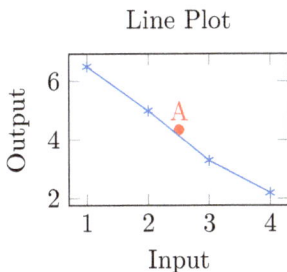

```
\begin{tikzpicture}
  \fill[red] circle(2pt)
    coordinate (A) node[above] {A};
  \begin{axis}[at=(A), anchor=center,
      scale only axis,width=3cm,height=2cm,
      title=Line Plot,
      xlabel=Input, ylabel=Output]
    \addplot[blue,mark=asterisk]
      table{data.dat};
  \end{axis}
\end{tikzpicture}
```

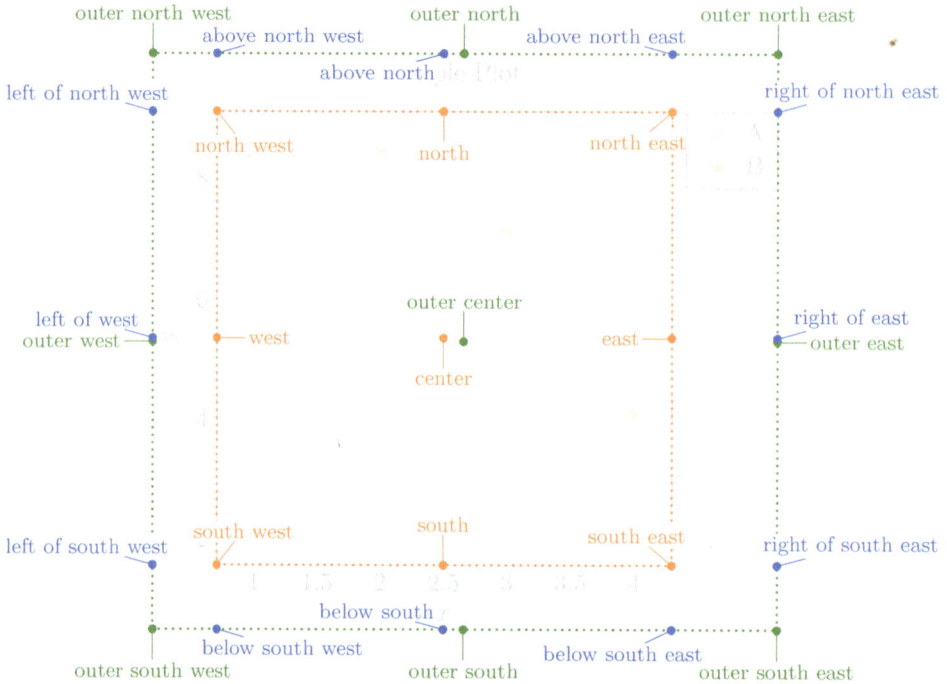

Figure 10.1: Pre-defined anchor points for `axis`. The orange dotted lines show the inner axes bounding box, and green dotted lines indicate the outer axes bounding box. Anchors on the inner axes bounding box are shown in orange, projections of those points onto the outer bounding box are shown in blue, and the compass points on the outer axes bounding box are shown in green.

The reference points on axes can be useful with group plots. A `groupplot` can be named using the key `group name=Name`, and then each axis within the group plot has a name of the form *Name cCrR* where *C* is the column index and *R* is the row index. For example, the coordinate (`foo c2r1.below south`) refers to the anchor `below south` of the plot in column 2 and row 1 of the group plot with `group name=foo`. If you want each subplot to have its own reference-able caption, then it is possible to use the `\subcaption` command from the `subcaption` package to produce subcaptions inside of nodes. It is important that the nodes containing the subcaptions have a specified `text width`, or the subcaption command will produce an error.

Example 10.9 shows an example of a LaTeX figure with a `groupplot`, showing how to include subcaptions. The LaTeX `subcaption` package provides the `\subcaption` command, and the LaTeX `float` packages provides the H option for the figure placement. Observe that we create a `subcap` style setting a fixed width (4 cm) and position `below`. We also set `inner ysep=0em` since the `\subcaption` command already adds appropriate vertical space. We can use the standard LaTeX `\label` command and then produce cross-references to the subfigures using the standard LaTeX reference, e.g., `\ref{fig:a}`.

Example 10.9: Group plot with external subcaptions

```
Preamble:\usepgfplotslibrary{groupplots}
Preamble:\usepackage{subcaption,float}
\tikzset{subcap/.style={below,text width=4cm,inner ysep=0em}}
\begin{figure}[H]
  \centering
  \begin{tikzpicture}
    \begin{groupplot}[
        group style={group name=foo, group size=2 by 1},
        scale only axis, width=4cm, height=1.5cm,
        xmin=0.5,xmax=4.5,xtick distance=1,
        ymin=0,ymax=12,ylabel near ticks ]
      \nextgroupplot \addplot[blue,mark=*] table[y=A]{data.dat};
      \nextgroupplot \addplot[red,mark=*] table[y=B]{data.dat};
    \end{groupplot}
    % Adding subcaptions
    \node[subcap] at (foo c1r1.below south)
      {\subcaption{Dataset A}\label{fig:a}};
    \node[subcap] at (foo c2r1.below south)
      {\subcaption{Dataset B}\label{fig:b}};
  \end{tikzpicture}
  \caption{Group plot with subcaptions}
  \label{fig:test}
\end{figure}

In Fig.~\ref{fig:test}, Dataset A is shown in Fig.~\ref{fig:a},
and Dataset B is shown in Fig.~\ref{fig:b}.
```

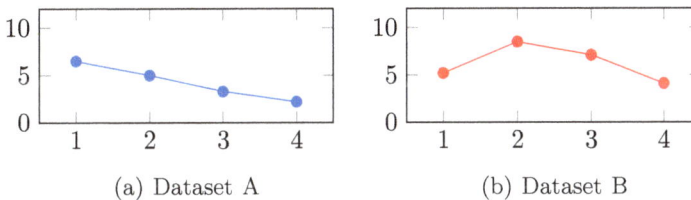

(a) Dataset A   (b) Dataset B

Figure 10.2: Group plot with subcaptions

In Fig. 10.2, Dataset A is shown in Fig. 10.2a, and Dataset B is shown in Fig. 10.2b.

10.6   Aligning multiple axes using anchors

From the discussion in the prior section, we see that axes can be placed according to various reference points. Thus, we can use that facility to align multiple axis nodes within a tikzpicture. (There is an alternative to groupplot discussed in Section 9.5; consider both!) In Example 10.10, we manually place four axes to display data from data.dat (on page 55). We size every axis to be 4 cm wide by 1.5 cm high. The first axis is named A, and so we can use its reference points to help in placing subsequent axes. We want to put the axis named B to its right. We create coordinate (BB) to the right of A.right of south east, adding a buffer of (0.5,0) for some extra space.

We then place axis name=B at that coordinate, anchored at B.left of south west. This place ensures that the x-axes of the top two plots are aligned. We want to place the axis named C below A. We create coordinate (CC) below A.below south west, adding a buffer of (0,-0.5). We then place axis name=C at that coordinate, anchored at C.above north west. This place ensures that the y-axes of the left two plots are aligned. Finally, we want to place the axis named D below B and to the right of C. We create coordinate (DD) so that it is aligned with B.west and C.north. We then place axis name=D at that coordinate, anchored at D.north west. This place ensures that the x-axes of the bottom two plots are aligned and the y-axes of the right two plots are aligned.

Example 10.10: Placing axes manually

```
\pgfplotsset{every axis/.style={scale only axis, width=4cm, height=1.5cm}}
\begin{tikzpicture}
  \begin{axis}[name=A,title=Dataset A]
    \addplot[blue,mark=square*] table[y=A]{data.dat};
  \end{axis}
  \path (A.right of south east) ++(0.5,0) coordinate (BB);
  \begin{axis}[name=B,anchor=left of south west,at=(BB),title=Dataset B]
    \addplot[red,mark=*] table[y=B]{data.dat};
  \end{axis}
  \path (A.below south west) ++(0,-0.5) coordinate (CC);
  \begin{axis}[name=C,anchor=above north west,at=(CC),title=Dataset C]
    \addplot[orange,mark=halfsquare*] table[y=C]{data.dat};
  \end{axis}
  \path (B.west |- C.south) coordinate (DD);
  \begin{axis}[anchor=south west,at=(DD),title=Dataset D]
    \addplot[teal,mark=pentagon*] table[y=D]{data.dat};
  \end{axis}
  % Show the coordinates (just for demonstration)
  \fill[gray] foreach \nd/\loc in {BB/below,CC/above,DD/below left} {
    (\nd) circle(2pt) node[\loc]{\nd}};
\end{tikzpicture}
```

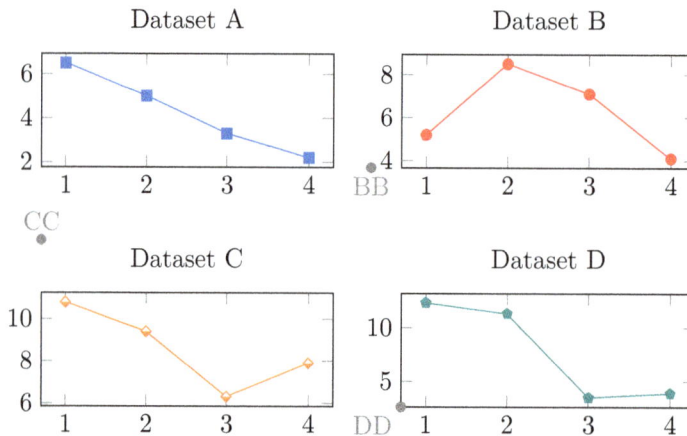

10.7 Detached legend

It can be useful to have one legend for multiple plots. The axis key `legend to name`=*Name* creates a standalone legend that can be placed separately. The legend is not drawn on the current axes, but it is instead saved so that it can be drawn later. (This is also a way to suppress a legend, since detaching it causes the legend not to be drawn.)

In Example 10.11, we demonstrate a detached legend and how it might be used when it is shared between two plots. Here we define two plot styles named `m1` and `m2`; we do this outside of the `groupplot` environment so the settings are still available when we actually draw the legend. We make the y-limits and many other settings the same for both axes by declaring these settings in the arguments to `groupplot`. In the first axis, we create a legend and save it using the `legend to name` option. It does not appear in the first axes. We plot the second axis and assign its legend to a name that we never use. Finally, we place the legend that has been saved using a node outside of the `groupplot` environment. We put it midway and below the two axes, specifically, midway between (`grp c1r1.below south east`) and (`grp c2r1.below south west`). The legend itself is recalled via `\pgfplotslegendfromname{sharedlegend}`. We put this inside a node for appropriate placement. This example uses `runtime.csv` (on page 68).

Example 10.11: Detached legend

```
Preamble:\usepgfplotslibrary{groupplots}
\begin{tikzpicture}
  \pgfplotsset{scale only axis,table/col sep=comma,
    m1/.style={orange,mark=diamond*,mark size=3pt},m2/.style={cyan,mark=*}}
  \begin{groupplot}[width=3cm, height=2cm,
      group style={group size=2 by 1, group name=grp, ylabels at=edge left},
      ylabel={runtime (sec)}, ymin=5, ymax=2e4, ymode=log,
      legend columns=-1, legend entries={Method 1~~, Method 2}]
    \nextgroupplot[title=CPU,legend to name={lgd}]
    \addplot[m1] table[y=M1]{runtime.csv};
    \addplot[m2] table[y=M2]{runtime.csv};
    \nextgroupplot[title=GPU,legend to name={lgd2}]
    \addplot[m1] table[y=M3]{runtime.csv};
    \addplot[m2] table[y=M4]{runtime.csv};
  \end{groupplot}
  \path (grp c1r1.below south east) -- (grp c2r1.below south west)
    node[midway, below] {\pgfplotslegendfromname{lgd}};
\end{tikzpicture}
```

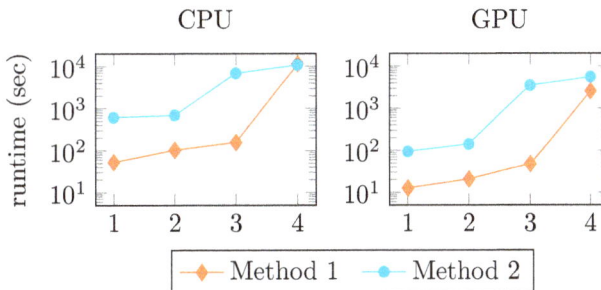

10.8   Nuances of plot marks and options

By default, the mark attributes (color, line style, etc.) are inherited from the plot options, but we can customize with the options in Reference 10.6.

Reference 10.6: Mark options

`\addplot[`*PlotOptions*`]` ...

Mark option	Description
mark=*Mark*	Specify mark shape; see Reference 7.4
mark size=*Size*	Set mark size
mark color=*Color*	Set secondary mark color; default is white

Style	Description
every mark	Modify mark colors, rotation, etc.

Some options are specified at the level of the plot options, such as mark size and mark color. Others have to be specified via the style every mark, such as draw and fill. Note that only marks with a * can be filled, such as square*. All the "half" marks such as halfdiamond* have two colors, except for halfcircle (the only half mark that has a non-asterisk version). The first color is the mark fill color (which defaults to the plot color) and the second is specified by mark color (which defaults to white). The exception is halfcircle, which has only the secondary color specified by mark color.

In Example 10.12, we show different ways of modifying the every mark style. Style A sets the fill color to be 20% of the plot color (blue). Style B changes the draw style of the mark to solid. Style C uses a two-color mark, halfcircle*, in the default color of black and a secondary color of yellow using mark color. It further rotates the mark by 45 degrees. This example uses data.dat (on page 55).

Example 10.12: Mark options

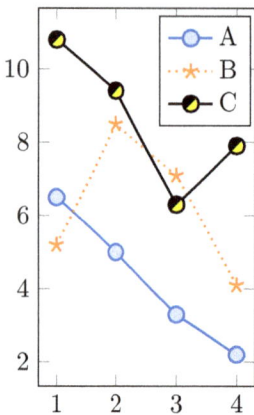

```
\pgfplotsset{scale only axis, width=3cm, height=5cm,
  every axis plot/.style={thick,mark size=3pt},
  A/.style={blue,mark=*,
    every mark/.append style={fill=.!20}},
  B/.style={orange,mark=star,dotted,
    every mark/.append style={solid}},
  C/.style={mark=halfcircle*,mark color=yellow,
    every mark/.append style={rotate=45}}}
\begin{tikzpicture}
  \begin{axis}[legend entries={A,B,C}]
    \addplot[A] table[y=A]{data.dat};
    \addplot[B] table[y=B]{data.dat};
    \addplot[C] table[y=C]{data.dat};
  \end{axis}
\end{tikzpicture}
```

> **?.?** **Oddity:** The option `mark color=Color` sets the *secondary* color for two-color marks such as `halfcircle*`. The primary mark colors default to the plot color and are changed via `every mark/.style={fill=Color, draw=Color}`.

10.9 Nuances of title style

The title node style is controlled by `every axis title`, which defaults to `{at={(0.5,1)}, above, yshift=6pt}`. The location for the `at` is with respect to the `axis description cs` coordinate system, meaning it is expressed as a percentage of the plot's width and height. Thus, the default sets the title at 50% of the width (centered) and 100% of the height. For that reason, the `yshift` default is 6 pt, which prevents the title from being rendered directly on the top axis. Note that you can override the `yshift` using `every axis title shift=Shift`. We recommend changing the default to 3 pt in Section 10.15. The key `title style` appends to the `every axis title` style. One thing that can be a bit tricky is that setting `yshift` inside of `title style` has an additive effect rather than overwriting the `yshift`. The location of the title can be modified via `at` and other node options.

> ⚠ **Remember:** Setting `yshift` inside of `title style` has an additive effect. Instead, use `every axis title shift` to provide an absolute *y*-shift.

10.10 Nuances of axis label styles

The keys `every axis x label` and `every axis y label` control the axis label styles, and you can append additional styles using `xlabel style` and `ylabel style`, respectively. By default, the *y*-label is a fixed distance of 35 pt left of the axis, and the *x*-label a fixed distance of 15 pt below the axis, independent of the width/height of the tick labels. The special keys `xlabel near ticks` and `ylabel near ticks` override the default distances and instead place the labels just outside the tick labels; these are recommended default settings (see Section 10.15). A nuance is that these special keys actually reset the corresponding label styles and so must be indicated before `xlabel style` or `ylabel style`. The locations of the axis labels can be modified using `at` and other node options.

> ⚠ **Remember:** The options `xlabel near ticks` and `ylabel near ticks` must be set *before* `xlabel style` and `ylabel style`.

10.11 Cycle lists

The default behavior of PGFPLOTS is to use something called a `cycle list` to set default plot styles. The idea of a `cycle list` is that each plot uses the next style in some predefined list, saving the user the trouble of explicitly declaring a distinct style for each plot. You can use an existing named cycle list (the default `color` is shown in Example 10.13) or define and use your own named cycle list. There is also `bar cycle list`, which is a set of default styles for bar graphs. I recommend disabling these via `cycle list={}` and `bar cycle list={}` in Section 10.15, and then specifying an explicit style for each `\addplot` command.

Example 10.13 illustrates the usage of the default cycle list. We add 13 plots where the nth plot is the constant function $y = n$. We reverse the y-axis so that the plots are going down in order. The style *cycles* with the 11th plot.

We recommend against using the `cycle list` feature because similar behavior can be achieved by using styles as in Section 7.7. The problem with cycle lists is that it is hard to jump around to select particular styles.

Example 10.13: Cycle list demonstration

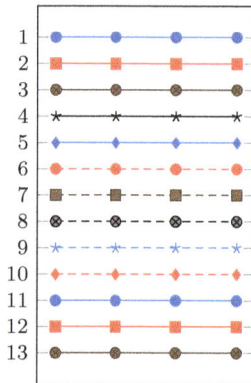

```
\begin{tikzpicture}
  \begin{axis}[
      scale only axis, width=3cm, height=5cm,
      font=\footnotesize, y dir=reverse,
      xtick=\empty, ytick={1,2,...,13},
      every axis plot/.style={samples=4},
      cycle list name=color, % PGFPLOTS default
    ]
    \foreach \n in {1,2,...,13} {
      \addplot expression {\n};
    }
  \end{axis}
\end{tikzpicture}
```

10.12   Nuances of axis width and height

It is perhaps counterintuitive, but the default for PGFPLOTS is that setting `width=W` and `height=H` yields axes of size $(W - 45\text{pt}) \times (H - 45\text{pt})$ as shown in the first axis in Example 10.14. The extra 45 pt in each dimension roughly accounts for the addition of a title, axis labels, and tick labels. Adding the option `scale only axis` changes the default behavior so that setting `width=W` and `height=H` results in axes of size $W \times H$, as shown in the second axis of Example 10.14. We strongly recommend making this `scale only axis` a default; see Section 10.15.

? *Oddity:* Setting `height=H` and `width=W` results in size $(W - 45\text{pt}) \times (H - 45\text{pt})$. Specify `scale only axis` so that the axis is instead $W \times H$.

Example 10.14: Effect of `scale only axis`

```
\tikzset{
    sizer/.style={orange,<->,dashed,thick},
    snode/.style={midway,sloped,auto,font=\sffamily},
    tnode/.style={above,align=flush center,xshift=3mm,font=\ttfamily,gray}}
\begin{tikzpicture}
  \begin{axis}[height=3cm, width=3cm, name=A]
    \addplot[blue,mark=*] table{data.dat};
  \end{axis}
  \draw[sizer] (A.left of south west) -- ++(90:3cm) node[snode] {3 cm};
  \draw[sizer] (A.below south west) -- ++(0:3cm) node[snode,below] {3 cm};
  \node[tnode] at (current bounding box.north) {height=3cm, width=3cm};
\end{tikzpicture}
```

```
\hspace{2cm}
\begin{tikzpicture}
  \begin{axis}[scale only axis, height=3cm, width=3cm, name=B]
    \addplot[blue,mark=*] table{data.dat};
  \end{axis}
  \draw[sizer] (B.left of south west) -- ++(90:3cm) node[snode] {3 cm};
  \draw[sizer] (B.below south west) -- ++(0:3cm) node[snode,below] {3 cm};
  \node[tnode] at (current bounding box.north) {scale only axis,\\
    height=3cm, width=3cm};
\end{tikzpicture}
```

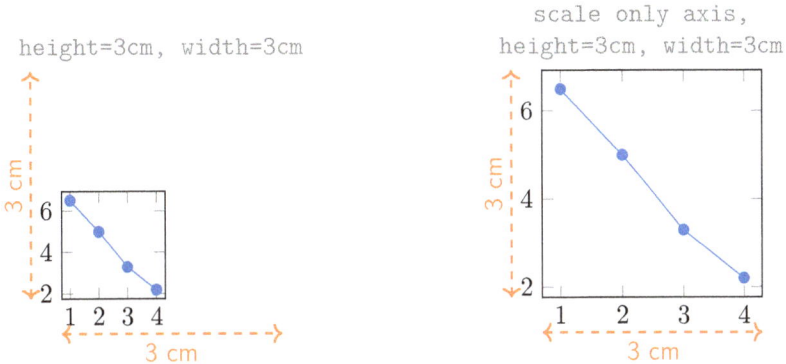

10.13   Mixing plot types

10.13.1   Scatter and line plot

It is possible to combine a scatter plot and a line plot. To make this work, you must specify [only marks, scatter, scatter src=explicit symbolic] as \addplot options rather than axis options. In Example 10.15, we first add the scatter plot and then a line plot. The plots and legends work as expected. This examples uses points.csv (on page 61) and data.dat (on page 55).

Example 10.15: Combined scatter and line plot

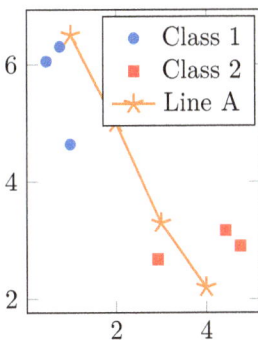

```
\begin{tikzpicture}
  \begin{axis}[
      scale only axis, width=3.2cm, height=4cm,
      title=Scatter \& Line Plot,
      legend entries={Class 1,Class 2,Line A} ]
    \addplot[only marks, scatter,
        scatter src=explicit symbolic,
        scatter/classes={
        a={blue,mark=*},b={red,mark=square*}}]
        table[meta=class,col sep=comma]{points.csv};
    \addplot[orange,thick,mark=star,mark size=4pt]
        table{data.dat};
  \end{axis}
\end{tikzpicture}
```

10.13.2   Line and bar plot

Combing a ybar and line plot is a bit more complex, especially if we want to display a legend for the two plots together. If we specify ybar as an axis option, we must override this effect when drawing our line plot. Conversely, if we specify ybar as an \addplot option, we must provide additional options in order for the legend to render correctly. We discuss these two variants below, both using data.dat (on page 55).

The first option, shown in Example 10.16, specifies ybar as an axis option. The first plot is automatically a bar plot. To make the second plot a line plot, we have to add sharp plot to the \addplot options. If there is a legend, then line legend has to be additionally added. If the line plot is first, the bars will plot over the lines but, interestingly, not over the marks.

Example 10.16: Combined ybar and line plot, version 1

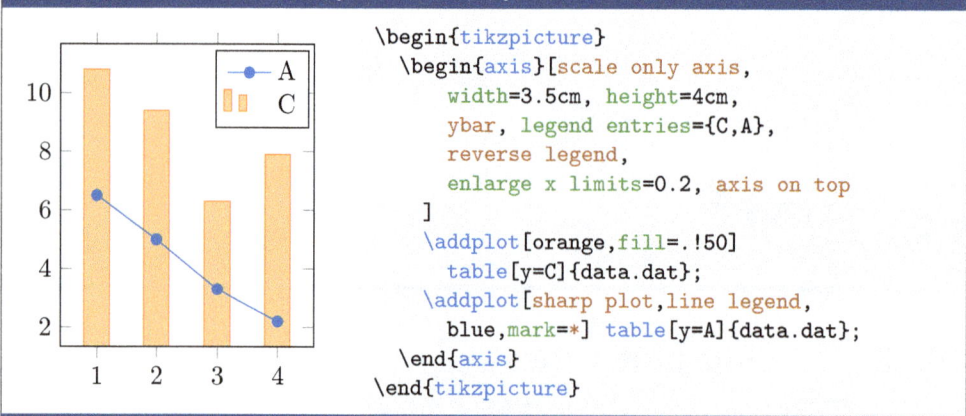

```
\begin{tikzpicture}
  \begin{axis}[scale only axis,
    width=3.5cm, height=4cm,
    ybar, legend entries={C,A},
    reverse legend,
    enlarge x limits=0.2, axis on top
  ]
  \addplot[orange,fill=.!50]
    table[y=C]{data.dat};
  \addplot[sharp plot,line legend,
    blue,mark=*] table[y=A]{data.dat};
  \end{axis}
\end{tikzpicture}
```

The second option, shown in Example 10.17, is to not specify ybar as an axis option. In this case, we specify ybar as an \addplot option. For the legend to be correct, we have to also add ybar legend to the \addplot options. Again, the bars must be printed first or they will overlap the lines, though not the marks, of the blue plot.

Example 10.17: Combined ybar and line plot, version 2

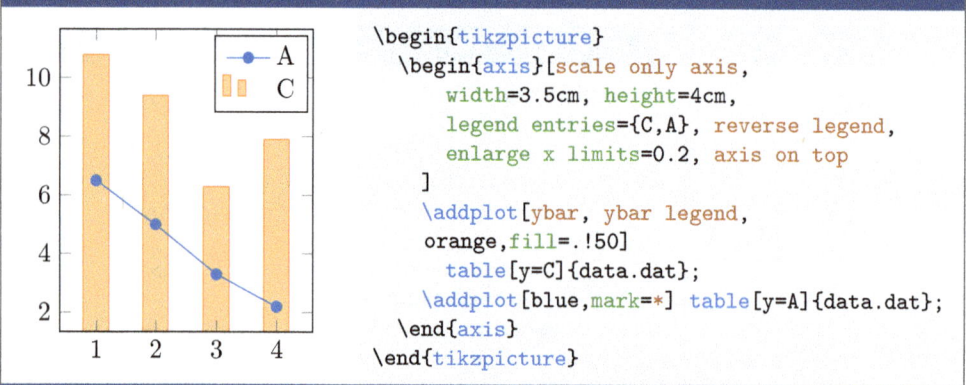

```
\begin{tikzpicture}
  \begin{axis}[scale only axis,
    width=3.5cm, height=4cm,
    legend entries={C,A}, reverse legend,
    enlarge x limits=0.2, axis on top
  ]
  \addplot[ybar, ybar legend,
    orange,fill=.!50]
    table[y=C]{data.dat};
  \addplot[blue,mark=*] table[y=A]{data.dat};
  \end{axis}
\end{tikzpicture}
```

Specifying the ybar plot type at the axis level automatically shifts the bars if plotting multiple bar plots and sets up the legend to be the bar symbols. If ybar is not at the axis level, the legend part is an easy fix (using the ybar legend), but the shifting has to be handled more delicately!

10.14   Caution with loops in pgfplots

Loops using \foreach may not work as expected inside the axis environment. In order to determine axis limits and other layout elements, PGFPLOTS delays evaluation of some expressions, meaning that the expansion of the loop variables may not happen as anticipated. Additionally, advanced loop options, like having two arguments, do not work properly inside an axis environment.

Example 10.18 demonstrates a workaround to make \foreach loops work inside an axis environment. The idea is to use the standard LATEX \edef to explicitly expand the definitions of loop variables combined with \noexpand to prevent expansion of all other macros. Here, we have a loop with two loop variables, \lval and \clr. We define an expanded macro \temp and then call it to immediately expand the loop variables. The other macros (\addplot, \addlegendentry, and \path) are protected from expanding at this point via \noexpand.

Example 10.18: Loop within PGFPLOTS

```
\begin{tikzpicture}
  \begin{axis} [scale only axis,
      width=8cm, height=4cm, font=\small,
      xtick=data, xmax=10, enlarge x limits=0.05, ymin=0, ymax=0.65
    ]
    \foreach \lval/\clr in {0.5/blue, 2.5/orange, 5.0/violet}
    {
      \edef\temp{
        \noexpand\addplot[\clr,very thin,mark=*,draw=black]
          expression[samples at={0,1,2,...,11}]
          {exp(-\lval)*(\lval^x)/factorial(x)};
        \noexpand\addlegendentry{Poisson $\lambda=\lval$}
        \noexpand\draw[\clr,dashed]
          (axis cs:\lval,0) -- (axis cs:\lval,0.65)
          node[midway,auto,sloped] {$\lambda=\lval$};
      }\temp
    }
  \end{axis}
\end{tikzpicture}
```

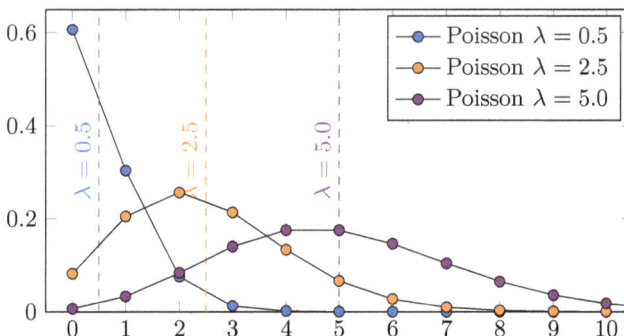

Inside a groupplot, it is even more difficult to have a loop. In this case, we must use \pgfplotsforeachungrouped and \eappto from the etoolbox package. The \eappto is like \edef but enables appending a number of expressions to the expanded macro before its evaluation. We give an example of this in Example 10.19, using data.dat (on page 55).

Example 10.19: Loop within groupplot

```
Preamble:\usepackage{etoolbox}
Preamble:\usepgfplotslibrary{groupplots}
\begin{tikzpicture}
  \begin{groupplot}[group style={group size=2 by 2},
      scale only axis, width=4cm, height=2cm,]
    \def\tmp{}
    \pgfplotsforeachungrouped \ltr in {A,B,C,D}{
      \eappto\tmp{
        \noexpand\nextgroupplot[title={Dataset \ltr}]
        \noexpand\addplot[blue,mark=*] table[y=\ltr]{data.dat};
      }}
    \tmp
  \end{groupplot}
\end{tikzpicture}
```

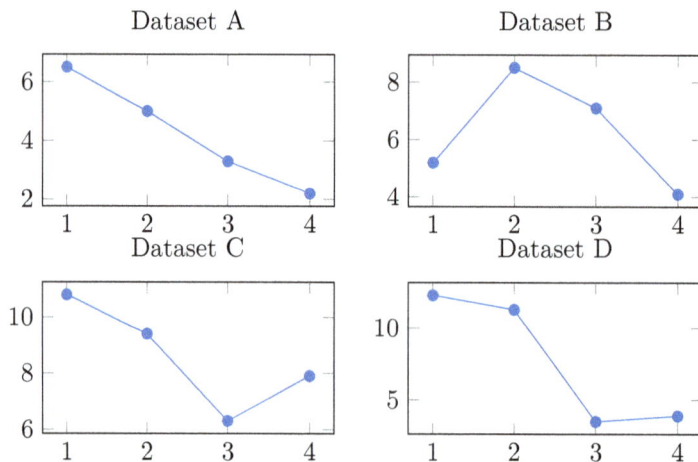

10.15   Recommended settings

One of the challenges of using PGFPLOTS is that some of the settings are difficult to modify and several defaults are non-intuitive. To ameliorate at least a few of these issues, I recommend the following settings in your preamble.

Reference 10.7: My recommended settings for PGFPLOTS

```
\pgfplotsset{
  every axis/.append style={
    scale only axis,
    xlabel near ticks, ylabel near ticks,
    every axis title shift=3pt,
    cycle list={},
    bar cycle list={},
  },
  every axis plot/.append style={
    every mark/.append style={solid},
  },
  numformat single/.style={/pgf/number format/#1},
  numformat/.style={numformat single/.list={#1}},
}
```

Adding the option scale only axis changes the default behavior so that setting width=W and height=H results in axes of size $W \times H$ versus $(W-45\,\mathrm{pt}) \times (H-45\,\mathrm{pt})$. See Section 10.12 for further details.

Similarly, the options xlabel near ticks, ylabel near ticks ensure the labels are placed dynamically depending on the size of the tick labels. See Section 10.10 for details.

The setting for every axis title shift is a matter of preference. The default of 6 pt seems too far to me, so I suggest 3 pt. See Section 10.9 for details.

Likewise, erasing the cycle list and bar cycle list prevents potential confusion in how the style of a plot is determined. It will need to be manually specified; see Section 10.11 for details.

The setting that specifies that every mark is solid is a repeat of Reference 7.5, which forces the mark line to be solid even if the plot line is dashed. More details about mark styles can be found in Section 10.8.

The user-defined numformat style is a repeat of Reference 9.4 and makes it easier to specify the number formatting options so that numformat={fixed, precision=2, zerofill} is a more compact way to express /pgf/number format/fixed, /pgf/number format/precision=2, /pgf/number format/zerofill. See Section 9.2 for more information.

Appendix A

LATEX lengths and colors

A quick reminder about colors and lengths in LATEX. We provide some useful units of length such as `pt` in Fig. A.1. The default step unit in TikZ pictures is 1cm. The default unit for most other options (e.g., `shorten <`) is in points (`pt`) if not specified.

pt	Point = 0.351 mm
mm	Millimeter = 2.845 pt
cm	Centimeter (default TikZ graph unit)
ex	Height of a small "x" in current font (approximate)
em	Width of capital "M" in current font (approximate)

Figure A.1: Length units in LATEX

The default available colors from the `xcolor` package are shown in Fig. A.2. These can be mixed, e.g., `green!70!black` produces ▭. Such colors can be saved and named via the `\colorlet{Name}{Combo}`; e.g., `\colorlet{mygreen}{green!70!black}`.

▭	black	▭	blue	▭	brown	▭	cyan
▭	darkgray	▭	gray	▭	green	▭	lightgray
▭	lime	▭	magenta	▭	olive	▭	orange
▭	pink	▭	purple	▭	red	▭	teal
▭	violet	▭	white	▭	yellow		

Figure A.2: Predefined colors in LATEX (using `xcolor` package)

Colors can be arbitrarily defined via `\definecolor{Name}{Model}{Values}`. Models include `gray`, `rgb`, `RGB`, `HTML`, and `cmky`. For example, the color ▭ can be defined any of the following equivalent ways:

- `\definecolor{mygreen}{HTML}{009B55}`
- `\definecolor{mygreen}{RGB}{0,155,85}`
- `\definecolor{mygreen}{rgb}{0,0.608,0.333}`
- `\definecolor{mygreen}{cmky}{100,0,45,39}`

Appendix B

Speeding up Compilation

B.1 Externalization

One downside to using TikZ is compilation speed, but TikZ has the ability to *externalize* all TikZ graphics so that they are only compiled after a change. This dramatically speeds up the compilation, but it requires some special setup.

In the preamble of your document, be sure to invoke the `external` library and enable externalization.

> **Reference B.1: Enable externalization**
>
> ```
> \usetikzlibrary{external}
> \tikzexternalize[prefix=DIRNAME/]
> ```

Externalization can be temporarily disabled via `\tikzexternaldisable` and reenabled via `\tikzexternalenable`. The `\tikzexternaldisable` command only applies to the current scope, e.g., inside a `figure` environment, so reenabling may not be needed.

Filenames are set automatically, but it can be useful to provide specific file names using `\tikzsetnextfilename`. The argument `\tikzsetfigurename` sets the prefix for all figures that follow, which can be useful if naming figures within a chapter.

> ⚠ **Remember:** Use `\tikzexternaldisable` while a graphic is still in development mode, making it easier to debug.

The trickiest part is that LATEX compilation requires the `-shell-escape` flag so that it can make its own system calls to compile the externalized graphics.

B.2 Precompilation of headers

It's also possible to combine the TikZ externalization with header precompilation, but this gets a bit complicated!

First, *all* of the frontmatter to be precompiled (starting with `\documentclass{...}`) goes into a file that we'll call `preamble.tex`, as follows.

> **Reference B.2: Example preamble file: `preamble.tex`**
>
> ```
> \documentclass{article}
> \usepackage{tikz}
> \usetikzlibrary{arrows.meta}
> \usetikzlibrary{topaths}
> \usetikzlibrary{positioning}
> \usetikzlibrary{shapes.geometric}
> \usetikzlibrary{calc}
> \usetikzlibrary{math}
> \usetikzlibrary{external}
> \usepackage{pgfplots}
> \usepgfplotslibrary{fillbetween}
> \usepgfplotslibrary{groupplots}
> ...
> ```

The preamble is precompiled into `preamble.fmt` using the command:

> **Reference B.3: Compiling preamble**
>
> ```
> pdflatex -ini -job-name="preamble" "&pdflatex preamble.tex\dump"
> ```

This must be run every time the file `preamble.tex` or its dependencies are modified.

There are two ways to tell LaTeX to use the precompiled preamble. The first is as an explicit argument to the compilation command, i.e.,

> **Reference B.4: Using precompiled header (Option 1)**
>
> ```
> pdflatex -fmt=preamble.fmt main.tex
> ```

The second is to make the first line of the file to be a special comment of the form:

`%&preamble`

with no spaces or other changes.

TikZ externalization plays nicely with precompilation, but there are a few caveats. First, you have to change the default TikZ externalization system call to include `-fmt=preamble.fmt`. Second, the `\tikzexternalize` command cannot be in the precompiled header. So, the resulting `main.tex` file would look like the following.

Reference B.5: Example `main.tex` using precompiled header

```
%&preamble

% Telling TikZ externalize how to find the precompiled header
\tikzset{external/system call={pdflatex -fmt=preamble.fmt
    -shell-escape \tikzexternalcheckshellescape -halt-on-error
    -interaction=batchmode -jobname "\image" "\texsource"}}

% Next command cannot be in precompiled header!
\tikzexternalize[prefix=extfigs/]

\begin{document}
...
\end{document}
```

Index

Boldface indicates primary pages. Italics indicate interesting examples.

www.ingramcontent.com/pod-product-compliance
Lightning Source LLC
Chambersburg PA
CBHW080242270326

41926CB00020B/4348